Schriftenreihe
der Wissenschaftlichen Landesakademie
für Niederösterreich

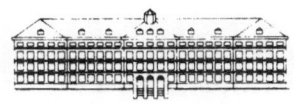

P. Kopacek (ed.)

Robotics in Alpe-Adria Region

Proceedings of the
2nd International Workshop (RAA '93),
June 1993, Krems, Austria

Springer-Verlag Wien New York

Dipl.-Ing. Dr. Peter Kopacek
Scientific Academy of Lower Austria, Krems, Austria

With 133 Figures

ISSN 0940-5801
ISBN-13: 978-3-211-82545-7   e-ISBN-13: 978-3-7091-9346-4
DOI: 10.1007/978-3-7091-9346-4

Preface

Industrial robots as a kind of a mechatronic system were the objects for intensive scientific research in the last years. Kinematics and kinetics, advanced control algorithms, flexible robots, mobile robots, cooperation of robots were research topics. Meanwhile the industrial robot is more or less a tool on the shop floor level like NC- and CNC-machines, transportation and storage devices. The current research landscape on industrial robots can be divided in two directions. The first direction is the scientific one and topics like fuzzy control, applications of neuronal networks, application of methods of artificial intelligence for robot control, optimal path planning are currently headlines in this field. On the other hand the application oriented research goes in the direction to develop and create new low-cost concepts including industrial robots applicable in a commercially efficient way mainly in small and medium sized companies.

The industry in most of the member countries of the Alpe-Adria Group are dominated by small and medium sized companies. Industrial robots together with the appropriate CIM-concepts are a very efficient tool for increasing the flexibility of such companies. At the first meeting in Portoroz (June 1992), a first overview on robotics research and applications in Alpe-Adria countries was given. First steps were done in the direction of a broader cooperation in science, development, production and level of education among these countries. This second meeting in Krems tried to establish new links between scientists and managers to develop priority suggestions for governments and investment strategies. Furthermore the participants should be informed on the goals of this group and proposals for international cooperation in terms of common projects should be discussed on the goals of this group.

For this second workshop in Krems more than 60 abstracts were submitted to the organizers. According to the scope and the topics the papers deal with review of industrial applications of robots in Alpe-Adria Region, review of research activities, review of university and research centres, low-cost automation for small and medium sized companies, robots and intelligent manufacturing system especially an assembly automation and last but not least education in robotics. In this proceedings selected papers are collected.

I hope that this workshop and the written material will stimulate the cooperation and definition of common research projects and activities in the field of robotics in the broadest sense in the Alpe-Adria Region.

P. Kopacek
Editor

Krems, December 1993

Contents

Robot Control

Sensors and Mobile Robots

Applications

Robots in Manufacturing Cells

Robots in Testing

Various Aspects

Steering Committee

G. Belforte   Politecnico di Torino, Italy
E. Fugger   Austrian Research Centre Seibersdorf, Austria
K. Jezernik   University of Maribor, Slovenia
P. Kopacek   Technical University of Vienna, Austria
J. Lenarcic   Jozef Stefan Institute Ljubljana, Slovenia
J. Marton   State Office for Technical Development, Hungary
A. Rovetta   Politecnico di Milano, Italy
U. Stanic   Jozef Stefan Institute Ljubljana, Slovenia

Organising Committee

P. Kopacek   Chairman
N. Stanek
A. Weichselbaum   Secretary
A. Kefeder   Secretary

Sponsored by

Austrian Ministry of Science and Technology
Scientific Academy of Lower Austria
Institute for Handling Devices and Robotics,
Technical University of Vienna

Robotics Research in Austria

K. Desoyer, P. Kopacek und I. Troch
University of Technology, Vienna, Austria

Abstract: This paper is divided in industrial applications and research at universities and research institutions. According to the last statistics at the end of 1992 approximately 1750 robots were in use in Austria. In the paper a short overview on the research on robotics is given. There are two main directions: More theoretical oriented research in the field of kinematics, kinetics, path planning and control and research mainly emphasized to applications mainly in small and medium sized companies. In the last time research is done in the direction of application of methods of artificial intelligence in form of expert or knowledge based systems as well as in fuzzy control and neuronal networks related to the field of robotics.

1. Robot statistics (Moos, 1991)

In Austria approximately 20 companies dealing with robots or robotic systems are on the market. In 1989 robots to the value of 300 Mio ATS were installed new in our country and 872 industrial robots were in use in Austrian companies. At the end of 1991 this number increased to 1465 and at the end of 1992 to approximately 1750. Therefore Austria is one of the countries with the highest growing rates according to the long term estimations because most of the other countries have smaller growing rates.

An usual characteristic number for the "robotization" of the industry of a distinct country is the number of robots for 10000 employees in this country. This number has increased from 3.4 in 1986 to 7.5 in 1989 and to approximately 10.0 today. One of the reasons for this untypical behaviour in comparison with other countries is the structure of the Austrian industry. The Austrian industry is mainly dominated by small and medium sized companies. These companies started the introduction of industrial robots relatively late - approximately 1984. Today 2 % of the industrial robots are installed in small companies (less than 20 employees); 49 % in medium sized companies (less than 500 employees).

The main application field of industrial robots in Austria is welding (22 %) and not spot welding (only 3 %) as usual in other countries. Handling operations on die-casting (17 %) and CNC machine tools (15 %) are also relevant application fields. Remarkable growing rates are obtained for assembly, palettizing, jet cutting, research and education applications. Today 35 industrial robots are installed on Universities and other research institutions.

2. Robotic research

As pointed out earlier the Austrian industry is mainly dominated by small and medium sized companies. Therefore robotic research on universities as well as research institutes have to be oriented mainly to industry. Because of the size of Austria these institutes are usually very small in comparison with other countries. We have two Technical Universities (Vienna and Graz) and one University with a Department of Technical and Natural Science (Linz). Therefore robotic research is mainly concentrated on these Universities and on Research Institute (ÖFZS, Seibersdorf). The well known sentence "small is beautiful" is valid for robotic research in Austria. The average size of research groups in Austria is approximately 4. This requires high flexibility and motivation of the researchers. "Theoretically" oriented research on robotics in Austria is financed by some scientific founds. The main task of the Austrian research should be the field of industrial applications. In the Austrian industry a strong demand for flexible and modular "low cost" solution for robotic systems can be obtained.

To transfer international knowledge in the Austrian industry some international scientific events have been or will be organized in Austria e.g. the

IFAC Symposium on "Theory of Robots", 1986, Vienna (Kopacek et al., 1986),

IFAC Symposium on "Robot Control", 1991, Vienna (Troch et al., 1991)

IFAC Symposium on "Low cost Automation", 1992, Vienna (Kopacek and Albertos, 1992),

IFAC Workshop "CIM for small and medium sized companies", 1992, Vienna (Kopacek, 1992, 1),

2nd International Workshop "Robotics in Alpe-Adria Region", 1993, Krems (Kopacek, 1993),

2nd IFAC Workshop on "Intelligent Manufacturing Systems", 1994, Vienna,

4th International Workshop "Robotics in Alpe - Adria Region", 1995, Krems,

International Workshop on "Social Aspects of Robots", 1995, Vienna.

Only one manufacturer of robots dealing with some research activities.

2.1 Research at Universities

In the following only some examples of University research activities in the field of robotics can be given.

In the Department of Mechanical Engineering (University of Technology of Vienna) the first Institute in Austria, specialized in manipulators and robotics, has been founded in January 1990. This Institute is responsible for the education of students mainly form Mechanical Engineering in robotics and related fields such as microsystems for external sensors, assembly oriented construction, software for robot control.

The institute is dealing with the field of robot-systems and systems of handling devices, using both theoretical and practical approaches (Kopacek, 1992, 2). The theoretical-oriented investigations are related to kinematics and kinetics of industrial robots, and the use of advanced control algorithms. Furthermore emphasis is placed on construction of light-weight robots, methods of pattern-recognition and application of methods of artificial intelligence.

Broad space is left to the use of industrial robots in Austrian industry - especially in small- and medium-sized companies - in connection with tasks of assembly technique. This is very often accomplished in the context of CIM-concepts. For that reason CIM-systems are also part of interest of the institute.

The task of the institute is to do theoretical research in the field of robotics, where industrial applications seem to be future oriented. Furthermore results in research should be utilized for Austrian industry. The companies of interest are small and medium sized companies. This leads to following research topics: kinematics and kinetics of industrial robots, advanced control algorithms, optimal path planning, flexible robots, co-ordination of robot motions, sensors and sensor-signals (microsystems), applications, robots in CIM-concepts, assembly with industrial robots, assembly-oriented construction, methods of artificial intelligence in robotics, social and cultural aspects of robotics and low-cost automation (Kopacek, 1992, 2).

In the next study term lectures on assembly automation, intelligent robots, advanced robot control and intelligent manufacturing systems will be given. The institute of robotics has closed contacts to the department of Systems Engineering and Automation at the Scientific Academy of Lower Austria in Krems.

In co-operation with the Department of "Systems Engineering and Automation" of the University of Linz an assembly cell equipped with 2 robots was planned, installed and tested in a medium sized company in Upper Austria (Kopacek and Fronius, 1989). This assembly cell works since April 1990 and serves for assembling of 6 different types of primary parts of welding transformers. It is integrated in a "low cost" CIM system (Kopacek et al. 1992) which is developed by the two Institutes mentioned above. In this project some peripheral devices such as a soldering device, special grippers and some storage units were developed. In the last time a knowledge based system for diagnosis and maintenance was installed in this cell.

The research of the interfacultary Working Group on "Robotics and Handling Devices" at the Technical University of Vienna is more theoretically oriented. Main topics of the research of this group are kinematics and kinetics of industrial robots, modelling, control and simulation of robots with new methods, optimal path planning in co-operation with the "Institute of Robotics", and collision avoidance. In the field of lightweight robots a new arm structure was designed with 40 % less of mass and the same stiffness than conventional arms. New algorithms of optimal path planning lead to faster movements and a higher accuracy.

At the "Division for Mathematics of Control and Simulation" at the University of Technology of Vienna problems of robot control are investigated. Path planning and robots with redundant DOFs are of primary interest. For movements along a geometrically given path, time-suboptimal algorithms could be improved considerably. Moreover, an algorithm was developed which yields the true time-optimal solution, is easy to handle and can be used on-line in certain cases. These algorithms were and still are tested for various examples in collaboration with the robotics group at ENSP Strasbourg. Work in progress is concerned with optimization of construction parameters for special purpose robots.

The division has a close co-operation with the "ARGE - Industrieroboter und Handhabungsgeräte" at the University of Technology Vienna and it was within this common project that Allgeuer wrote his thesis (Allgeuer, 1991), which lead to a promotion sub auspicis präsidentis. He dealt with robots with redundant DOFs and investigated the question under which conditions, repeated movement along a given geometric path yields identical movements of the robot's joints, a question being important e.g. with respect to efficiency and security (collision avoidance). Application of modern mathematical methods yields criteria for repeatability of a redundant robot as well as methods which were successfully tested for several examples of practical relevance.

At the time being, algorithms are developed for robots with redundant DOFs which guarantee that the hand point moves in such a manner along a geometrically given path that no joint or any part of a link has a collision with objects close to this path. It is expected that one of these algorithms under development can be used on-line.

The needs of the Austrian industry led to the following research topics: Collection and preparation of the geometric, kinematic and kinetic fundamentals for drives, arms and grippers for various types of robots used today. Investigation of possibilities for constructive improvements with the goal to minimize computational and control effort as well as construction costs; catalogization of dynamic models for the different parts (e.g. arms, gears, actuators) and combination of them to models for the whole robot. Development of simple models which are accurate enough for the application of observers. Development of models which take the elastic deformation into consideration, and simplification of such models. The latter can be performed in many cases by simulation only. Collection and development of "advanced", digital control algorithms for the position control. Theoretical and practical comparison of them with regard to suitability and efficiency. Collection and comparison of various methods for path calculation, especially for optimal paths. Simulation tests of control algorithms with the aid of the models being developed. Test of selected non-linear control algorithms. Test of selected "observers". Application of sensor signals in order to increase the efficiency of the control.

The Institute of "Flexible Automation" (INFA) is working in a very broad range of topics in the field of manufacturing and automation and in the field of robotics (Zeichen, 1992). Research subjects in robotics are: robot metrology, robot calibration, modular robot controlling systems, fuzzy technology in robot control, digital signal processors.

Knowledge of a robots dynamic behaviour and of the absolute accuracy characteristics turns out to be most useful in the development as well as application of robots. Nowadays there are no appropriate measurement systems available to acquire this knowledge. Thus a measurement system to dynamically track the end-effector of a robot was successfully developed.

A laser beam is deflected over a gimbal mounted mirror into a retroreflector which is mounted to the robot's end-effector. The retroreflector parallel reflects the beam. Due to the motion of the robot the beam will not strike the retroreflector in its center, thus there is a parallel displacement between the incoming and the reflected laser beam. This offset is measured and used by the tracking controller to follow any movement of the robot.

A system for robot calibration makes non contact, dynamic measurements with high accuracy in a large working area and in real-time. Since the measurement process in automatic, it is a very flexible system that enables several applications besides the measurement of robot characteristics.

One of these applications is to use the measurement data, especially pose or path accuracy data, to calibrate the parameters of the robot model. It has been shown that the overall accuracy of a robot can be improved from 3 millimeters to 0.1 millimeters. Besides the measurement accuracy this improvement depends on the model capabilities. Depending on the specific robot, incorporating factors such as backlash, deflection in the joints or links, temperature sensitivity or tumbling of the joint axes increases the number of parameters but also the level of description and therefore the robot accuracy. The developed robot model includes all this factors in a non redundant and singularity free description. With this accurate model robot accuracy can be improved towards the order of robot repeatability.

Some introductory work has been done on the topics of object-oriented robot simulation. Together with the Minsk Radioengineering Institute, USSR a C++ class library has been programmed. Using this class library a high-level task-oriented programming of robot tasks is possible. This class library could be the basis of a knowledge based robot task planning and simulation system.

The Institute of "Manufacturing Technology" in the Department of Mechanical Engineering of the University of Technology of Vienna was responsible for the foundation of an interuniversitary CIM Center called IUCCIM. In this center 8 University Institutes from the University of Technology of Vienna, from the University of Economics in Vienna as well as some companies are involved. In this center industrial robots are installed mainly for tasks of assembly automation as well as machine loading and unloading. Research activities are concentrated to use industrial robots in CIM concepts.

At the Institute of "Mechanics" at the Technical University of Graz some research is done in the field of kinematics of industrial robots. Research work is now concentrated to replace one arm of a robot by one with a new kinematic structure.

The Institute of Production Technology at the Graz University of Technology is highly specialized in precision engineering and metrology.

In the field of robotics a new Anthropoidic Measuring Device AMG-1 has been developed, designed and built at IFT (Kovac and Frank, 1993). On one hand AMG-1 can be seen as a passive robot, without drives and control but containing resolvers of highest resolution and accuracy, on the other hand AMG-1 is a co-ordinate measuring machine with rotatory joints. When connected to a robot, AMG-1 can be used for testing the accuracy of the robot as well as a measuring device, which is handled by the robot. A new mechanic device has been developed which makes it possible to measure geometric errors as well as the dynamic behaviour of a robot. Both the pose axes and the orientation axes are covered.

Furthermore a universe Test Field for Robot accuracy has been installed: With the new calibration device the home position and the length and position of all joints of a robot can be determined precisely by means of a simple mechanical equipment consisting of marketable elements.

The Department of "Systems Engineering and Automation" at the Scientific Academy of Lower Austria in Krems (Kopacek, 1992,3) is involved in research projects dealing with the development of a flexible, modular and "low cost" robotized assembly cell and a sensory equipped low cost gripper for robots (Kopacek and Fronius, 1989). In this project researchers from Slovenija are involved.

The planning of assembly cells was done by hand until now. In regard to the increasing amount of cells this does not seem to be economical in near future. Goal of this project is the development of a software, based on a relational database, which allows a computer aided planning of an assembly cell. In the first step a description of the parts of the assembled products in necessary. After that the operations are described by means of a specially developed "Assembly Symbol Language", which allows a standardized and formalized description of the operations. The components necessary for the realization of this assembly task are determined under direction of the planner on base of the task description (parts, operations), the data of the single components available in the component database. The utilization of the planning software should decrease the time of passage in planning of flexible assembly cells and increase the quality at the same time.

Since 1993 a 3 year research project with the laboratory of assembly automation at the University of Ljubljana dealing with intelligent assembly is supported by the Austrian government.

2.2 Research in Centers

Two research Institutes in Austria work in the field of robotics besides of other projects. The Austrian Research Center Seibersdorf (ÖFZS) is mainly involved in practically oriented research. Currently projects are: robots in assembly automation, for palletizing and for welding operations.

RISC-Linz (Research Institute for Symbolic Computation) deals with symbolic computation covers all algorithmic aspects of solving problems with symbolic (i.e. non-numeric) objects (Kopacek, 1992, 3).

One of the most promising areas of applications for symbolic computation is soft-automation including robotics. Many sub problems in the simulation, analysis, control, and supervision of robots or whole robot working cells can be attacked by symbolic techniques like geometric modelling, computational geometry, algebraic geometry, knowledge engineering, computer graphics. Research topics in past and current projects are: Gröbner bases for computing inverse kinematics, path planning for non-synchronized motions, Roider method for collision detection, NC-SAVE, simulation and verification environment for NC machining, Voronoi diagrams for path finding, robot vision based on geometric modelling.

2.3 Research by Industry

The main producer of industrial robots in Austria, the company IGM, has developed a robot controller based on transputers during the last years. Transputer technology offers the possibility to build up an open and modular control concept. Furthermore in that company a laser based sensor was worked out primarily for welding operations. This sensor should be used in the nearest future for flame cutting operations as well as for assembling operations.

For new applications serve two examples. Three robots in identical compact systems weld passenger car seats. For welding the seat frames and the back-rests a very restricted area is available. Loading and clamping of the single components, and unloading of the completed parts is done manually. For the manufacturing in 2 stations, identical simple devices are used. Shortest possible cycle times require optimum workpiece accessibility in the holding fixtures and minimum loading times.

The single part of the heat exchangers are manually put into the fixture. For the subms.sequent soldering the fixing brackets are to be accurately positioned on the workpiece. For this purpose, the robot effects up to 10 tack welds per bracket, needing 2 seconds per tacking point.

3. Summary

Austria as a very small country has only a real chance in the field of robotics - like in other so called high tech fields - by trying to modify and applicate research results mainly for small and medium sized companies. This requires the development of flexible, modular low cost solutions for robotic systems.

In the near future assembly automation will be of great importance for the Austrian industry. Therefore the development of low cost sensors in the direction of intelligent robots must be a research topic in Austria in the future probably by means of tools from microsystems.

Furthermore the methods of artificial intelligence e.g. intelligent diagnosis systems should be introduced in robotics research in Austria.

4. Bibliography

Allgeuer, H. (1992). Zur kinematischen Steuerung von Robotern mit redundanten Freiheits-graden. Doctor-Thesis, TU Vienna (Adv. by I. Troch and K.Desoyer).

Desoyer, K., Kopacek, P., Troch, I. (1991). Activity report of the working group on Industrial Robots and Handling Devices", Univ. of Technology Vienna.

Kopacek, P. (Ed.) (1992, 1). Preprints of the International Workshop "CIM for small and medium sized companies", Vienna.

Kopacek, P. (1992, 2). Activity report of the Institute of "Handling devices and Robotics", Univ. of Technology Vienna.

Kopacek, P. (1992, 3). Activity report of the Department of "Systems Engineering and Automation", Scientific Academy of Lower Austria, Krems.

Kopacek, P. (Ed.) (1993). Preprints of the 2nd International Workshop "Robotics in Alpe-Adria Region", Krems.

Kopacek, P., Albertos, P. (Eds.) (1992). Preprints of the International Symposium on "Low cost Automation", Vienna.

Kopacek, P., Fronius, K. (1989). CIM Concept for the Production of Welding Transformers. Preprints of the IFAC Symposium "Information Control Problems in Manufacturing Technology" (INCOM '89), Madrid, Volume 2, p. 737-740.

Kopacek, P., Frotschnig, A., Zauner, M. (1992). CIM for small companies. Preprints of the IFAC Workshop on "Automatic Control for Quality and Productivity-ACQP '92" Istanbul, p. 35-41.

Kopacek, P., Troch, I., Desoyer, K. (Eds.) (1986). Preprints of the International Symposium on "Theory of Robots", Vienna.

Kovac, I., Frank, A. (1992). A novel Industrial Robot Calibration Device. Proceedings of the 1st International Workshop "Robots in Alpe-Adria Region", Portoroz, p. 162-167.

Kovac, I., Frank, A. (1993). Robot guided Anthropoidic Measuring Device. Preprints of the 2nd International Workshop on "Robots in Alpe-Adria Region", Krems, p. MO 1.1-1 - MO 1.1-5.

Moos, H. (1991). Industrieroboterstatistik. WIFI, Vienna.

Stifter, S. (1993). Research Activities in Robotics at RISC-Linz. Preprints of the 2nd International Workshop "Robotics in Alpe-Adria Region", Krems.

Troch, I., Desoyer, K., Kopacek, P. (1991). Preprints of the International Symposium on "Robot Control", Vienna.

Zeichen, G. (1992). Activity report of the Institute for Flexible Automation, Univ. of Technology Vienna.

Considerations on Objectives, Tasks, and Organisation of Alpe-Adria Centre for Robotics and Automation

J. Lenarcic, U. Stanic, P. M. Oblak
Jozef Stefan Institute
Ljubljana, Slovenia

Abstract: The paper presents the initiative of constituting an Alpe-Adria Centre for Robotics and Automation that would integrate research, development, and educational laboratories in the region of Alpe-Adria. The purpose of the Centre is to facilitate the exchange of research and development achievements, students, faculty, and experts, as well as collaboration in common research programmes.

1. Introduction

Robotics has been around for some thirty years. Expectations about the capabilities of robots have always been described very favourable, but the available robots are very poor duplicates of man and can only be used for very straightforward jobs of a strictly repetitive nature. Any kind of human-like intelligence is still far out of sight and will be for quite some time to come. However, there is no doubt that robotics will develop very fast in the direction often dreamt of. The benefits of robot introduction can be seen on both micro and macro scales, for a given company and for the nation. The following can be easily recognised:

- The flexibility and versatility of industrial robots make possible the automation of multiproduct small-batch and mixed-flow line production. A complete changeover or even a modification in a product model often requires changing or at least radically rebuilding a special-purpose automated machine. Where a robot is used instead, a mere change in program is all that is required. As the product life cycle shortens, the flexibility and the versatility of robots become increasingly advantageous.
- Unlike man, robots can operate on a 24-hour basis. Robots are capable of performing functions at high speeds or loads that exceed human limitations.
- The sustained stability of robot operation, and their ability to work continuously and accurately provides a smoother production flow.
- The introduction of robots changes the nature of the production system. The characteristics of robots, combined with this change in the production system, has led to the creation of completely new technologies.

In the recent years, crucial political and economical changes resulting in a decreased amount of investments have affected the development and introduction of robots. Nevertheless, the wide socio-economic impact of the applications of robots in industry, such as improved productivity, stability and quality, production management, humanisation of working life, and resource conservation, have begun to be evident. The world robot population is, therefore, still increasing every year. Robotics is a conglomerate of various disciplines , such as mechanics, electronics, control, computer science, artificial intelligence, and is strongly connected with national economies. As an interdisciplinary activity, robotics calls for international co-operation. In many countries and communities, large-scale international projects of research and development have been launched that involve universities and research institutions, as well as companies ventured into manufacture of robots and peripheral devices.

After a period characterised by very optimistic results in research and market forecasts, the tendency now is to take more realistic and pragmatic approach. In connection, high level effort in research and development not only reinforces and improve the expertise, but also prepare the robotics for tomorrow. That should be the purpose of any national and international initiative. Traditional cultural and technological connections between nations in Alpe-Adria region has always represented an advantage. Robotics, however, is an area of practical interest that still has to be developed. It is a general agreement that a co-operation in technological and educational level would contribute to a more effective exploitation of all national resources.

The objective of the this paper is to present the initiative of constituting an Alpe-Adria Centre for Robotics and Automation that would integrate the research and development laboratories around Alpe-Adria. This centre would promote, stimulate and recommend international co-operation in robotics research, development and education and would particularly emphasise the comparative advantages of these traditionally connected nations, their socio-economic, cultural, and other relations. The initiative for the Alpe-Adria Robotics and Automation Centre should not interfere with other international initiatives of bilateral or multi-lateral co-operation. We trust that it should represent a new form or aspect of this co-operation for which we do expect support of national governments.

2. General Concept of the Centre

The Alpe-Adria Robotics and Automation Centre would be constituted by and from the existing research, development, and educational laboratories spread in the region of Alpe-Adria. The purpose of the Centre is to facilitate (or to enable) faculty and students, as well as other specialists, from Alpe-Adria countries and specialising in different disciplines to work together on basic and applied research projects in numerous state-of-the-art laboratories. The Centre should conduct research in robotics technologies relevant to for Alpe-Adria region and enable the adoption of the results of the research by industry. Three basic activities of the Centre are foreseen: research, development, and education.

Research - The current generation of robots have shown themselves to be useful for a variety of tasks, yet many of the most interesting robot applications in such fields of industry, civil engineering, and agriculture are beyond their scope. Such tasks include dextrous manipulation, automatic assembly, and vehicle guidance in unstructured environments. All these tasks have in common the need of sophisticated robot mechanisms with incorporated sensors as an integral part of robot control cycle. This requires a great deal of research, basic and applied, which cannot be carried out in all its complexity by one even extremely well equipped laboratory. Much more effective is a distributed research of more specialised laboratories with co-ordinated and complementary research programs.

Development - The easiest way for a company to be informed on the recent advances in the field of robotics and to attract trained people is to collaborate with researchers and other companies in a specialised research and educational institution. The existing laboratories and research groups have already established projects to permit the affiliated companies access to research results and to maintain liaison with projects of interest to them. By constituting the Alpe-Adria Robotics and Automation Centre, the companies would get access to a wider variety of programs and would have information on the latest results in different countries. This would stimulate a more productive dissemination of the most recent investigations in industrial environment and would, consequently, improve the responsiveness and competition of companies in the Alpe-Adria region, as well as stimulate them for a more substantial financial commitment.

Education - Industrial needs over the next decade require bright, exceptionally capable, engineering specialists, possessing initiative, drive, and a strategic perception of their role. To match the demands of industry and national and international development strategies, educational activities must be organised with a strong connection to research and development activities. The Centre would provide a wide variety of educational programs and would enable exchange of faculty and students in accordance to their demands. An international education centre of this type would meet higher educational standards and effectiveness than, for instance, a traditional university system that, for all its many strengths, is still largely organised around its traditional origins and institutional structure with too much inertia.

The goals of the Alpe-Adria Robotics and Automation Centre are the following:
- to co-ordinate national programmes in the field of robotics and automation
- to recommend and stimulate a wider international co-operation
- to recommend and stimulate technological and scientific exchange
- to contribute to the standardisation
- to facilitate the transfer of research results
- to guarantee transfer of automation technologies
- to educate potential users of robotics technology and to train specialists
- to represent the involved laboratories in national and international institutions

The Alpe-Adria Robotics and Automation Centre would be organised in terms of an international project. This project should not be in contradiction with any other international institution, bilateral or multi-lateral co-operation. It still has to be decided, however, how to make operate the project, either as a part of the existing schemes of international co-operation, such as the "Middle Europe Initiative", or as a unique initiative proposed by all participants directly to their national governments.

3. Economic Impacts

Undoubtedly, robotics has made a significant impact on the economic growth of the most developed countries, Japan, USA, Germany, Sweden. In the 1980s, industrial robots became the symbol for the whole area of factory automation. The number of industrial robots in a country or in industrial branch was often taken as a proxy of the technological level of that country or branch. Many governments set up massive programmes consisting of grants, or tax incentives, or the promotion of industrial robots. On the other hand, how many underdeveloped countries have considered a development strategy that includes automation in general and robotics in particular? We believe, however, that robotics represents a unique occasion for the underdeveloped to advance their technology, and thereby create economic links with more developed world. In contrast to the developed countries, the process of introduction of robots in underdeveloped countries is still very slow. Many economic and social factors contribute to this, such as the need of high investments, organisation of education, international economic co-operation, a technologically advances environment, public initiative, and dedicated funds to help programs on Industrial, and international levels. The world-wide economic recession in 1990s has made the robotics development more difficult even in the most developed countries, and practically impossible in less developed countries. Governments, industries, as well as other economic subjects, are looking for a more effective technological development that would take advantage of all possible national and international resources. The constitution of the Alpe-Adria Centre for Robotics and Automation would represent and opportunity for establishing scientific, technological and economical links between different countries in the region where so many things are connecting these nations but where very controversy political organisation in the last 50 years has incapacitated more intensive collaboration.

From the technological viewpoint, the development of robotics in the recent years has been characterised by:

- a trend for a massive utilisation of robotics in new applications in industry in a wide variety of new jobs, and
- a trend for massive utilisation of robotics in new non-industrial and non-manufacturing applications.

The automotive industry is the foremost user of robots, the second greatest is the mechanical industry. The share of both is declining. Loading and unloading is now the most common application followed by welding, where arc-welding still has the smaller portion. Everyone is convinced that assembly is to become in the near future the most important application area, although we must recognise that its share currently remains relatively small. Robots are currently being introduced in many new areas, for instance, in medical applications such as patient care or surgery, and housekeeping. All this requires extensive theoretical and technological research that calls for international collaboration.

Another fact must be mentioned in the philosophy of introduction of robots. Many were convinced, in the initial decades of robotics, that it will develop in a direction of unmanned factory. According to this, manufacturers invested enormous amount of money to introduce robotics and other automation technologies in huge production areas. Now is has become clear that the future of robotics is in "symbiosis" with man. Robots are implemented in small and inexpensive work cells forming small islands of automation spread around the factory. This enables also smaller manufacturers to utilise the robotics technology in their production process, giving them more flexibility and adaptability for the requests of the market. Here is another opportunity of small nations in the Alpe-Adria region to, based on the proposed initiative, technologically reinforce their traditionally small companies and thus increase their productivity, quality, and competitiveness. Smaller companies, however, need help from the governments and other institutions in their investments, but they also need support from various technical and technological aspects, since they do not possess their own expertise in robotics.

4. Scientific and Technological Background

The programme of international collaboration in the Alpe-Adria region should have three main goals:

- to reinforce technological development toward product development and industrial diffusion,
- to structure the scientific and technical community for strengthening industrial development, and
- to identify the comparative advantages of the nations involved, as well as areas of research and development that should be emphasised in this region.

To achieve these basic goals, different actions must be implemented:

- the first action aims at deepening research in advances information-processing technologies (e.g. artificial intelligence, data bases, modelling and simulation, communication networks) for integration of various components such as CAD/CAM, production planning, maintenance, and control and programming,

- the second action aims a promoting technological advances by sharing public and industrial research on finalised projects concerning computer-aided manufacturing, industrial robots, non-manufacturing robots. Projects should be initiated on various topics, such as new generation of industrial, mobile, agricultural, and civil engineering robots, integration of flexible automation cells, CAD tools in process design, and optimisation,
- the third action aims a accelerating diffusion of basic technologies in mature industries. This goal is achieved through industry-specific projects which associate laboratories, technical centres, and companies. Various new industries should be attracted such as wood industry, food industry, textile and garment industry, as well as tourism.

The structure of the scientific and technical communities represents an important topic in the discussion of robotics. Different objectives have to be defined, for instance the creation of regional but coordinated centres of expertise that are intended to be focal points of excellence, where laboratories, teaching bodies, research teams, and business people share their expertise.

5. On Tasks, Manpower, and Financial Needs of the Centre

As an international institute, the Centre would represent a platform for international collaboration in terms of research, development, and educational programmes that operates on the following basis:

- the Centre is formed as an integration of the existing R&D laboratories in the Alpe-Adria region and exploits manpower, equipment and other resources of these laboratories,
- each of the involved laboratories operates as an independent subject and cares about its own financial means, development strategy, management, etc.,
- bilateral or multi-lateral collaboration is carried out based on the principle of reciprocity and in accordance to the existing schemes of international scientific and technological collaboration,
- the Centre is a subject of coordination, stimulation, and information in the fields of research, development, and education and operates in the sense of public institution,
- the Centre intervenes at local governments and other national and international institutions in order to get eventual support for the participation of the involved laboratories,
- details on operation, tasks, bodies etc. are defined in the Constitution of the Centre.

6. Bibliography

Lenarčič, J. (1990). Centre for Robotics and Automation. *High Technology Park* (N.H. Afgan, Ed.) Hemisphere Publishing, New York (USA)

Albus, J. (1988). The Golden Age That Could Be. *Proc. 19th Int. Symp. on Industrial Robots*, Sydney (Australia)

Aron, P. (1983). The Robot Scene in Japan. *Daiwa Securities America Report*, USA

Laffaille, A. (1986). Robotics in France: Trends and Overview, *Proc. 16th Int. Symp. on Industrial Robots*, Brussels (Belgium)

Engelberger, J.F. (1988). Domesticating the Industrial Robot. *Proc. 19th Int. Symp. on Industrial Robots*, Sydney (Australia)

Karlsson, J.M. (1991). *A Decade of Robotics*. Mekanforbundets Forlag - International Federation of Robotics , Stockholm (Sweden)

Advanced Control Concepts for Industrial Robots

P. Kopacek
Institute for Handling Devices and Robotics
University of Technology, Vienna, Austria
P. Otto, J. Wernstedt
Department for Automation and Systems Engineering
University of Technology, Ilmenau, Federal Republic of Germany

Abstract: Many tasks in robot control are very nonlinear and complex. These tasks can be performed in an advantageous way by using fuzzy methods or artificial neuronal networks. In the first part of this paper the optimal design of fuzzy controllers acting as common nonlinear time discrete controllers is investigated. In the second part artificial neuronal networks in connection with several controller structures are presented. Best results are obtained by using the structure of the adaptive model based robot control system. It is shown that the fuzzy controller and the artificial neuronal networks lead to better results than conventional control concepts.

1 Introduction

In recent years many new principles in the field of robotic control have been created and tested. A special charateristic of these developments are problems resulting by the control of dynamically coupled nonlinear systems under real time conditions.

The problem of robot control includes the planning of stable and robust algorithms for the coordination of the robot's arms movement during the following of special paths in the workspace.

The movement control usually is based on the definition of the desired movement of the robotic arms's joints axes, which leads to the planned roboter's arms motion. This requires the mathematical evaluation of the forces, needed to drive the joints to achieve the desired position and movement along the path (Fu et al. 1987).

The equations to describe the motion are quite complex and highly nonlinear. As additional difficulty there are nonlinearities as result of friction and various boundaries that are still neglected in many applications.

By using new methods of control theory and computer science typical structures for intelligent robotic control systems could be created (Tsafestas et al. 1991).

The architecture of these control circuits is depicted in Fig. 1.

This paper confines to application of fuzzy based methods and artificial neuronal networks in robotic control.

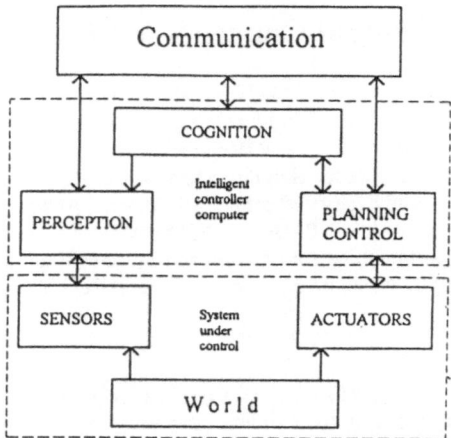

Figure 1: Typical structure for intelligent robotic control system

Up to now most of the commercially used industrial robots ignore the nonlinear characteristics mentioned before. This often leads to oscillations and overshoot in the positioning of the robot.

The continuously growing demands on the productivity and quality lead in increasing number to the usage of control methods that take care of the nonlinear problems.

The led to the creation of so called "model based controls" which improve the robots behaviour and pay attention to the dynamical charateristics (Tourassis 1988).

There are two major problems that stand in the way to the implementation of these methods:

1.   The mathematical evaluation of complex dynamical models has to be solved in real time (a complete cycle must only take some milliseconds).

2.   The model's parameters have to be known exactly, because otherwise an instable behaviour can occur easily.

Inaccuracies can result e.g. from model- or parameter deviations or disturbances.

Adaptive control algorithms reached a point, where they are able to take care of parameter variations and not aquisated dynamical characteristics (Aström 1983) but they are only usabel in linear systems. Today most interest concentrates on adaptive algorithms that are capable of tuning the dynamical parameters continuously and have the ability to compensate model deviations. Such methods use a predictor to approximate and compensate the model's aberrations (Tourassis 1987).

Another possibility is the usage of a nonlinear control algorithm that improves the handling of problems with modelling (Tourassis 1985).

A new direction in the development of robot control is the usage of fuzzy based algorithms or the implementation of artificial neuronal networks (Poo et al. 1992). The advantage of this knowledge based methods lies in:
* high parallelity
* ability to learn
* robustness against disturbances
* good results at nonlinear systems.

2 Robot control based on fuzzy methods

The advantage with the use of fuzzy control structures in robotics lies in a better performance in nonlinear processes. The fuzzy controller is used as common nonlinear discrete controller. It consists of some basic modules depicted in Fig. 2 (Kosko 1992).

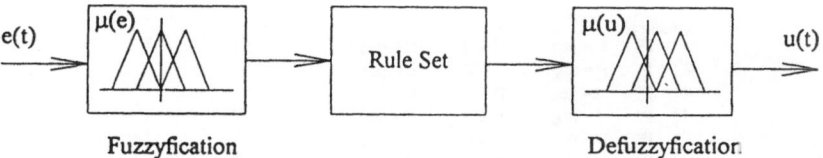

Figure 2: Structure of a Fuzzy Controller

To get an optimal design, the following assumptions are useful (Koch et al. 1993).

Step I: fuzzification

1. The membership functions have to be partially linear and to be described by up to four grid points.

2. For the membership function $\mu_i(e)$ is:

$$\sum_{i=1}^{n} \mu_i(e) = 1$$

Step II: defuzzification

1. As membership functions for the output $\mu(u)$ singletones are used

2. For the membership function $\mu_i(u)$ is:

$$\sum_{i=1}^{n} \mu_i(u) = 1$$

3. The computation of the output signal is done with min-max method by evaluation of the center of gravity.

In Fig. 3 the structure, membership function and the rule set are depicted.

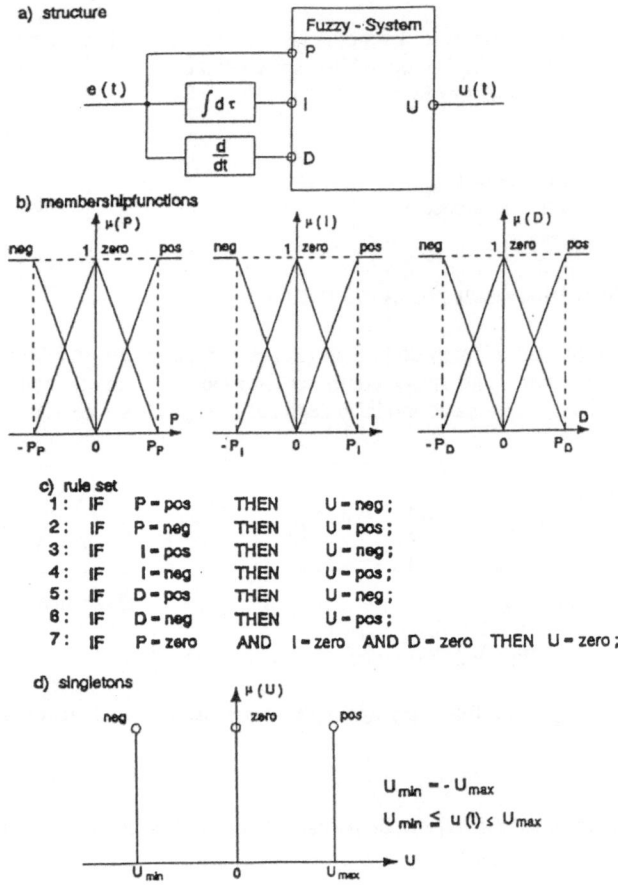

Figure 3: The elements of the simple fuzzy PID controller

Koch et al. 1993 could prove that an optimal design of a fuzzy controller is made possible only by the above mentioned restrictions. Further it is shown that there is no quality reduction caused by the drastical reduction in degrees of freedom in design.

The main problem with the design of fuzzy concepts for robots is the optimisation of the fuzzy controller.

In most cases this means a heuristic tuning of the fuzzy controllers by "Trial and Error". Resulting from nonlinearcontrol theory we try to find a possibility for a real optimisation of fuzzy controllers and the exact formulation of the optimisation task.

In figure 4 the multicriterial character of the optimisation of only one setpoint change is shown. But the equation shows the transformation of the multicriterial problem to a singlecriterical one by using the method of the weigth sum of particular criteria.

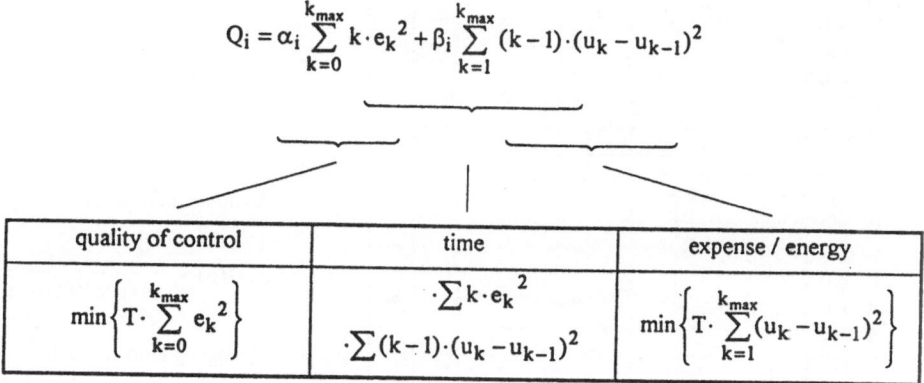

$$Q_i = \alpha_i \sum_{k=0}^{k_{max}} k \cdot e_k^2 + \beta_i \sum_{k=1}^{k_{max}} (k-1) \cdot (u_k - u_{k-1})^2$$

quality of control	time	expense / energy
$\min\left\{ T \cdot \sum_{k=0}^{k_{max}} e_k^2 \right\}$	$\cdot \sum k \cdot e_k^2$ $\cdot \sum (k-1) \cdot (u_k - u_{k-1})^2$	$\min\left\{ T \cdot \sum_{k=1}^{k_{max}} (u_k - u_{k-1})^2 \right\}$

Figure 4: Multicriterial character of the optimisation

While a linear control loop, optimised for a single setpoint change, is optimal in the whole linear workspace, the nonlinear control loop is only optimal for the one optimized setpoint change. Thus the design of fuzzy controllers can only be done for n various setpoint changes.

level 1: n - setpoints / changes

$$Q_{ges} = \sum_{i=1}^{n} \lambda_i Q_i$$

λ_i weight of set point i

with $\sum_{i=1}^{n} \lambda_i = 1$

level 2: setpoint i / change i

$$Q_i = \alpha_i \sum_{k=0}^{k_{max}} k \cdot e_k^2 + \beta_i \sum_{k=1}^{k_{max}} (k-1) \cdot (u_k - u_{k-1})^2$$

α_i weight of quality of control i

β_i weight of expense of control i

Fig. 5 shows the optimal design with the tool ILMFUZZY (Kuhn et al. 1992) for a typical nonlinear system. It is visible that the fuzzy controller is superior to the PID controller (Koch et al. 1993).

A comparison between PID and fuzzy controllers for position control has been done in experiments already in 1985 by Mandic et al. on a 6R/1000 robot. Mandic could reach very

18

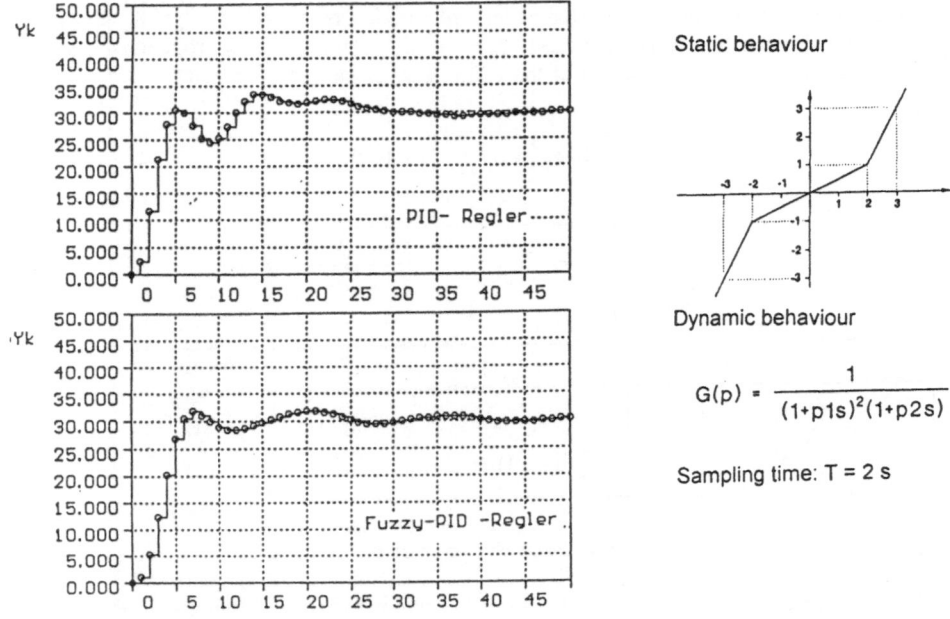

Figure 5: Comparison of PID and Fuzzy PID controller

good results using the nonlinear control concept depicted in Fig. 6. This shows the way for usage of fuzzy controllers in robotics.

Further tasks lie mainly in the design of controller structures for progressive fuzzy concepts and the creation of methods for the optimisation of control quality, energy consumption and time.

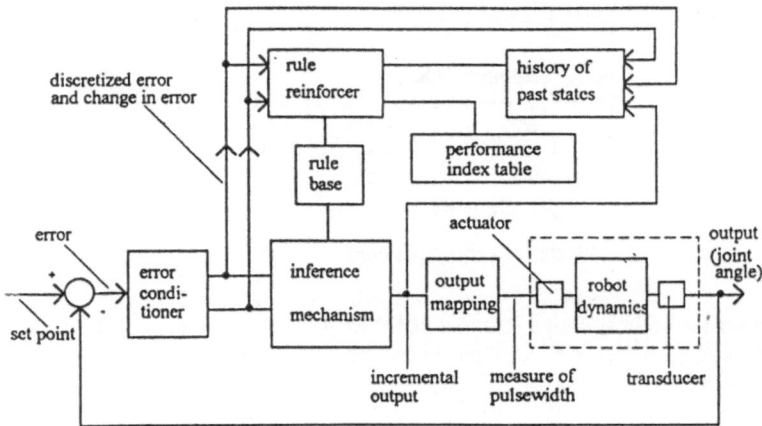

Figure 6: The fuzzy self organized robot system

3 Robot control based on artificial neuronal networks

Every artificial neuronal network application (ANN) in robot control contains the learning process of the dynamical behaviour of the robot (Poo et al. 1992, Miller et al. 1990).

The main differences lie in the form of embedding of the ANN in the control circuit, during teaching and in the adaptation.

1. Replacement of a conventional controller by an ANN (Fig. 7)

The ANN is trained off-line under usage of the in- and output of a conventional controller. The advantage lies in the fact of the higher speed of the ANN which improves the real time capabilities.

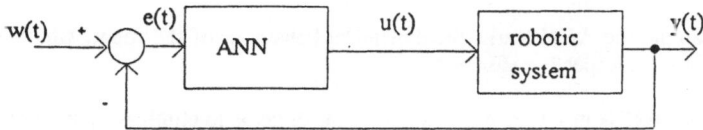

Figure 7: Conventional use of an ANN for robot control

2. Usage of the ANN in feed forward circuits

The ANN is trained on the dynamical behaviour of the robot and it is embedded in the control circuit as shown in Fig. 8 (Miller et al. 1987).

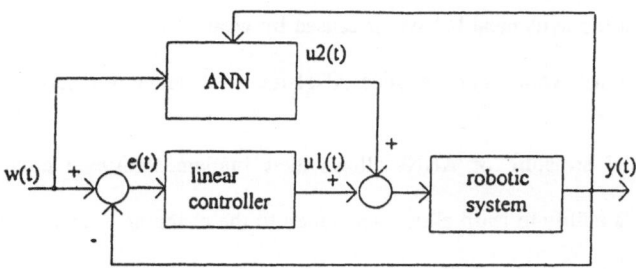

Figure 8: ANN acts as feedforward loop

The inputs of the ANN are the actual states of the robot and the needed minimum change of the current state to reach the desired path.

The output of the ANN corrects the force signal given by the controller to act at the corresponding link. The ANN is trained continuously in a working range for which the linear controller is designed. The ANN influences the robot in a way to improve accuracy by covering ranges which are badly covered by the linear controller.

3. Usage of an ANN in model based controls

The ANN is used to evaluate the nonlinear dynamical connections inside the model based control algorithm (Poo et al. 1992). The ANN is included in the forward branch of the control circuit and is used to compensate the nonlinear dynamical behaviour of the system so that it acts as linear system for the controller (Fig. 9).

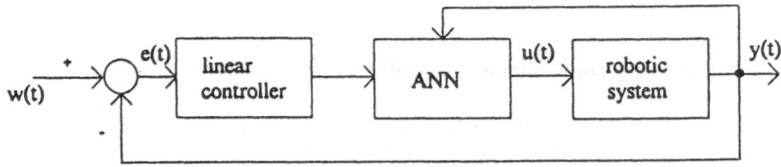

Figure 9: ANN used as linearizing adaptive controller

At each sample the ANN learns the dynamical behaviour of the robot, which enables it to tune itself to the respective situation.

The usage of ANNs in model based robotic control helps to eliminate some problems which occur by the implementation of conventional controllers.

These problems are:

*   The need to make simplifications in respect to the complex calculations of a dynamical model by using a closed loop control circuit.

*   A model that reproduces the whole dynamical behaviour of a robotic system does not exist (e.g. friction, loose, hysteresis).

*   Changes in the dynamical behaviour caused by wear.

Because of the following characteristics ANN are well suited for solving the problems mentioned above:

*   The parallel structure of ANNs allows their implementation in complex real time systems
*   The ANN's ability to learn allows adaptation to the changing dynamical behaviour of a robot.
*   The ability to recognise the important part of an information allows ANNs to make the robot control unsensitive against disturbances and measuring errors.

Although there are many different types of ANNs the "multylayer feed forward" networks which use the back propagation algorithm to learn are often used. They are able to describe the nonlinear relations between position, velocity, acceleration and the drive forces of the different joints (Fig. 10).

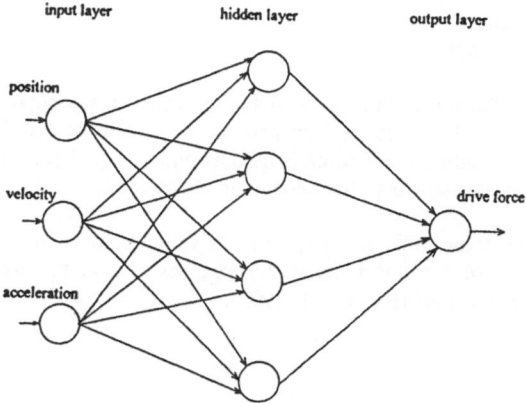

Figure 10: ANN used for the description of the robot's dynamical behaviour

Every node in the hidden layer uses a sigmoide decision function. The ANN is being prepared before its implementation in the model based control algorithm by a simulation model with nominal dynamical parameters. The fully trained ANN is then integrated in the robot control system.

During the phase of learning the inputs of the ANN consist of the signals for position $x(t)$ the velocity $\dot{x}(t)$ and the acceleration $\ddot{x}(t)$ evaluated from this terms instead of the control signal $u(t)$. According to the desired goal the reference signal $r(t)$ for the desired motion is being calculated. From this signal and the actual state of the robot the acceleration $u(t)$ is evaluated and used in combination with the actual state of the robot to serve as input signal for the ANN, which generates the needed force $F(t)$ (Fig. 11).

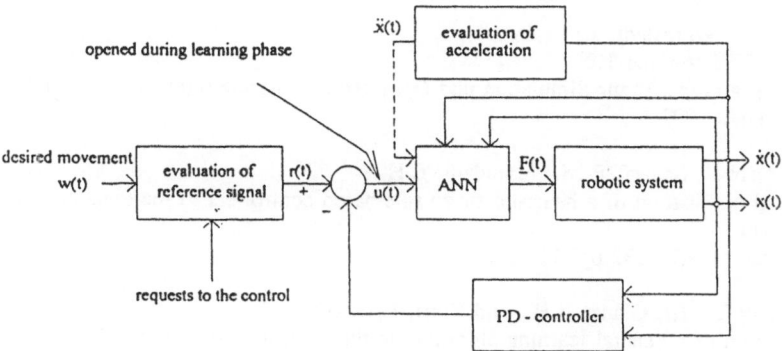

Figure 11: ANN based adaptive robot control

u(t) - acceleration
x(t) - position
$\dot{x}$(t) - velocity
$\ddot{x}$(t) - acceleration (evaluated)
r(t) - reference signal
$\underline{F}$(t) - Joint drive forces

This force drives the robot into a new state, for which the acceleration for the used drive force is calculated . This acceleration and the new state of the robot are used as input signals, the corresponding drive force is used as output signal for a data set for learning. In this way the ANN is adapted to the behaviour of the robot.

Experiments have shown that the usage of ANNs leads to good results in adaptive model based robotic control systems even if disturbances act on the robotic system or wrong measuring results are used (Poo et al. 1992).

References

Aström, K.J.:
Theory and applications of adaptive control: a survey.
Automatica, 1987, 19 (5), pp. 471-486

Fu, K.S., Gonzales, R.C. and Lee, C.S.G.:
Robotics: control,   vision and intelligence.
McGraw Hill, New York, 1987

Koch, M., Kuhn, Th., Wernstedt, J.:
Eine neue Entwurfsmethode für Fuzzy Regelungen.
at 41 (1993) H. 5, S. 152-158

Kosko, B.:
Neural Networks and Fuzzy Systems.
Prentice Hall, Inc. 1992

Kuhn, Th., Wernstedt, J.:
ILMFUZZY Version 1.0
Ein Programmpaket zur Simulation und Optimierung einschleifiger Fuzzy Regelkreise.
TU Ilmenau 1992

Mandic, N.J., Scharf, E.M., Mamdani, E.H.:
Practical application of a heuristic fuzzy rule-based controllers to the dynamic control of a robot arm.
Proc. IEE (1985), 132 pp. 190-203

Miller, W.T., III, Glanz, F.H., and Kraft, L.G. III:
Application of a general learning algorithm to the control of robotic manipulators.
Int. J. Robotics Res., 1987, 6 (2), pp. 84-98

Miller, W.T., III, Hewes, R.P., Glanz, F. H., and Kraft, L.G. III:
Real-time dynamic control of an industrial manipulator using a neural-network-based learning controller.
IEEE Trans., 1990, RA-6 (1), pp. 1-9

Poo, A.N., Ang Jr. M.H., Teo, C.L. and Qing Li:
Performance of a neuro-model-based robot controller: adaptability and noise rejection.
Intelligent Systems Engineering, Autumn 1992, pp. 50-62

Tourassis, V.D.:
Computer-control of robotic manipulators using predictors.
Proc. 1987 IEEE Symp. on Intelligent control, Philadelphia,
Pennsylvania 18-20 January 1987, pp. 204-209

Tourassis, V.D. and Neumann, C.P.:
Robust nonlinear feedback control for robot manipulatrs.
IEEE Proc. D, 1985, 132, pp. 134143

Tourassis, V.D.:
Principles and design of model-based controllers.
Int. J. Control, 1988, 47 (5) pp. 1267-1275

Tzafestas, S.G.:
Intelligent Robotic Systems.
Marcel Dekker, Inc., New York, 1991

IMProvement in Automation and Control Technology
A Joint European initiative

P. Gabko, M. Stierle
University Extension Centre
University of Technology, Vienna, Austria

Abstract: In the present paper the organisational and project management procedures of the TEMPUS JEP "IMPACT - IMProvement in Automation and Control Technology" is presented. The University Extension Centre, Vienna University of Technology, developed an interactive system oriented method for strategic technology forecasting. This method is strongly based on target oriented moderated discussions. The first part of the paper presents how this procedure was adapted to the generation of international projects, i. e. to TEMPUS projects. The second part of the paper is devoted to the presentation of the organisational structure, the management and first results of the JEP IMPACT. The conclusions drawn from the experience of the JEP IMPACT are the following. In the project the most important factors for success were seen in a detailed preparation of the JEP and the development of a "common understanding" of the aims and objectives by all JEP partners in the preparatory phase of the project.

1   Introduction

In summer 1990, the SEFI Task Force for Eastern Europe initiated to establish 5 SEFI TEMPUS Working Groups, in priority areas identified by the Eastern-Central European partners. The priority was given because of the strategic importance of the areas for the economic development of the eligible countries and because of the relevance for the institutional development of the partner universities from the eligible countries. The 5 priority areas were the following:

* automation and control
* biomedical engineering
* environmental engineering
* information technology, and
* management and business administration for engineers.

The University Extension Centre, Vienna University of Technology (UEC/VUT), committed itself to co-ordinate the SEFI TEMPUS Working Group on Automation and Control. During the last ten years, the UEC/VUT developed an interactive system oriented method for strategic technology forecasting (M.Horvat, R.Wimmer). This method is strongly based on target oriented moderated discussions between members representing different social systems.

Since the UEC/VUT is running a large project on "East-West Cooperation through Technology Transfer and Continuing Education" the aim with respect to the SEFI TEMPUS Working Group on Automation and Control was to adapt the above mentioned procedure to the generation of TEMPUS projects. Thus, this paper will present a review on the project generation and design procedure of the TEMPUS JEP "IMPACT-IMProvement in Automation and Control

Technology" in the first part and it will give some details on the management of the JEP in the second part.

2   Towards A Moderated, Target Oriented, Interactive Process

Before going into the details of the project generation and design some aspects of the methodology of the procedure are presented. In the point of view of the SEFI Task Force for Eastern Europe the TEMPUS programme only will fulfil its task if the projects really fit to the objectives and strategies for further development of the Eastern-Central European institutions involved (set M.Horvat, M.Stierle). Thus, special attention was paid to the process of project generation.

The project generation was seen as a complex communication process with different social systems (institutions) involved. The communication in the different universities from the eligible countries and the communication between the different institutions involved in the project were seen as the key elements of this complex process.

The aim of the procedure was the following:

* to clarify the context i.e. what is feasible in the framework of the TEMPUS Programme
* to identify the state of the art and the plans for further development of the partners from the eligible countries involved,
* to specify the priority subject fields of the broad area "automation and control"
* to identify appropriate EC and EFTA project partners
* to match offers and demands, and
* to prepare the JEP application.e was

The interactive character of the procedurguarantied through two meetings of a "Core Group" which consisted in the first meeting of 1 representative from every Eastern-Central European university respectively and 2 representatives from institutions of the SEFI Task Force for Eastern Europe (Vienna University of Technology and Helsinki University of Technology). In the second meeting the group was enlarged by representatives from other EC universities.

The communication in the Eastern-Central European universities involved was stimulated through the task of clarifying the needs and the future perspectives in the field of automation and control in a self evaluation process.

3   Project generation, project organisation and first results

Preparing the application

The SEFI TEMPUS Working Group on Automation and Control started its concrete activities in January 1991. Since scientists from different backgrounds participated in the project generation it was the task of the chairman to moderate a target oriented debate. The most important issues in this starting period were:

* to specify the priority subject fields on the basis of the state of the art of the broad area "automation and control technology" in the partner organisations in the eligible countries
* to identify the needs for further development at the Eastern-Central European partner institutions,
* to invite additional Western partners to participate in the project,
* to clarify which kind of activities are eligible in the TEMPUS programme, and
* to identify priority target groups for the planned JEP activities.

On the basis of all these issues, a project application was submitted to the EC TEMPUS Office in March 1991. In August 1991, the contract for the TEMPUS JEP "IMPACT-IMProvement in Automation and Control Technology" was signed.

Organisational structure

The organisational structure chosen for the JEP IMPACT was designed as decentralised as possible with the co-ordinating institution in a monitoring and supporting role.

The task of the contacting and co-ordinating institution (one half time co-ordinating assistant) is to distribute the finances by means of subcontracts, to co-ordinate and monitor the whole project and support concrete activities if a need is stated.

One half time co-ordinating assistant in every institutions from the eligible countries is in charge of organising and administering all aspects of the project concerning their institution.

The steering committee with one representative from every institution is responsible for the project strategy, for the budget and the contents of all activities.

Main activities in the first project year

The main activities in the first project year are the following:

* 7 jointly organised seminars and workshops
    + Workshop "University Industry Cooperation in Automation and Control Technology", Vienna
    + Workshop "Databases and Knowledge Systems", Miskolc
    + Workshop "Model Based Fault Diagnosis", Warsaw
    + Seminar on Automation and Control Technology Education "EXACT '92", Dresden
    + Workshop "Computer Aided Education in Automation and Control", Prague
    + Workshop "Economic Aspects of Automation", Wroclaw
* installation of equipment at 4 departments
* 2 steering committee meetings
* 11 short visits for academic staff (less than 2 weeks),
* 3 retraining mobilities for academic staff
* 1 teaching assignment
* 2 student study periods
* 6 student placements.

First results from the activities in the first project year

The main outputs have been achieved in the following two areas during the first project year:

    * upgrading of students' laboratories and
    * developing teaching materials, introducing new or upgrading existing courses.

Concrete results in upgrading of laboratories and practical exercises were identified in the following fields:

    * PLC Network programming (Miskolc),
    * Electronic Systems (Prague),
    * Computer Controlled Systems (Prague),
    * Modern Control Systems (Prague),
    * Laboratory of Automation (Warsaw),
    * Computer Applications Laboratory (Warsaw).

Concrete results in developing teaching materials, introducing new or upgrading existing courses were identified in the following fields:

    * Artificial Intelligence (Miskolc),
    * Database Systems (Miskolc),
    * "Process Control" (Miskolc),
    * Simulation of Systems (Prague),
    * System Identification (Prague),
    * Process Visualisation and Modelling (Prague),
    * Model Based Fault Detection (Prague),
    * Electronic Systems (Prague),
    * Computer Controlled Systems (Prague),
    * Modern Control Tools (Prague),
    * Computer Science (Warsaw),
    * Control Theory (Warsaw),
    * Discrete Processes Automation (Warsaw),
    * Automation (Warsaw),
    * MRP Application (Wroclaw),
    * Simulation Modelling in Management (Wroclaw),
    * production planing systems (Wroclaw).

Main activities in the second project year

The main activities in the second project year are the following:

    * 9 jointly organised seminars and workshops
        + Round Table discussion in the framework of the IFAC Symposium on "Low Cost Automation '92", Vienna
        + Summerschool on "Binary Control '92", Prague
        + Workshop on "Automation and Control Technology Education 2001", Vienna
        + Preparation of a book on "Control Engineering Solutions: A practical approach"
        + Workshop Parallel Processing in Education, Miskolc

+ Seminar on Automation and Control Technology Education EXACT '93,
  Dresden
+ Workshop on "Manufacturing Resource Planning", Vienna
+ Workshop "CIM Education", Vienna
+ International Workshop on Advanced Approaches in Industrial Control Systems,
  Prague
+ Summerschool Intelligent Systems for Control, Toulouse

* installation of equipment at 7 departments
* 3 steering committee meetings
* 14 short visits for academic staff (less than 2 weeks),
* 21 retraining mobilities for academic staff
* 3 teaching assignment
* 2 student study periods
* 4 student placements.

4 Conclusions

Three factors are seen as most important for the success of the project.

The first factor is seen in the detailed step-wise preparation of the JEP objectives really responding to the needs of the Eastern-Central European partners and the specification of the outcomes for every JEP activity in advance.

The second one is seen in creating a "common understanding" of the aims and objectives and respective outcomes to be achieved by all JEP partners in the preparatory phase of the project.

In close connection with the second aspect, a clear distribution of the roles in the project and the development of a strong commitment to the project by all project partners is very important.

References

Horvat, M., Stierle, M. (1992). The Approach of SEFI Task Force for Eastern Europe towards TEMPUS. In: Kveton, K. (Ed.). Proceedings of the International Conference on Trans-European Cooperation in Engineering Education, 18 ff.

Horvat, M., Wimmer, R. (1985). Wissenschaftstransfer, Report prepared for the Austrian Ministry for Science and Research

Stierle, M. (1991). A Procedure for the Development of TEMPUS Applications. Proceedings of the Conference Go East/Go West, Linz

Horvat, M., Stierle, M. (1992). TEMPUS JEP IMPACT - Project Generation, Design and Management. In: Kveton, K. (Ed.). Proceedings of the International Conference Trans-European Cooperation in Engineering Education, 264 ff.

Research Activities in Robotics at RISC-Linz

S.Stifter
RISC-Linz
Johannes Kepler University, Linz, Austria

Abstract: For quite a few years, robotics and, more generally, softautomation has become one of the main research topics at RISC-Linz (Research Institute for Symbolic Computation). Conditional on the fields of expertise of RISC-Linz, the concentration is on geometric problems like collision detection, collision free path finding, path tracking, forward and inverse kinematics, and software technological aspects as they appear e.g. in simulation systems, supervision systems, and control systems. Both, basic research as well as applications within industrial cooperations are studied.

1 The Institute RISC-Linz

One of the most promising areas of applications for symbolic computation is softautomation including robotics. Many subproblems in the simulation, analysis, control, and supervision of robots or whole robot working cells can be attacked by symbolic techniques like geometric modeling, computational geometry, algebraic geometry, knowledge engineering, computer graphics. The activities of RISC-Linz in this area range from basic research to cooperations with industrial partners in which software systems for different applications within robotics are developed. Research topics in past and current projects are: Gröbner bases for computing inverse kinematics, supervision system for tunnel constructions, path planning for non-synchronized motions, Roider method for collision detection, NC-SAVE, simulation and verification environment NC machining, Voronoi diagrams for path finding, robot vision based on geometric modeling, geometry theorem proving and constraint checking, supervision system for a window production line, 5-axes milling of spectacular frames, automated nesting of irregular shapes, milling tool constructions.

2 Selected Research Topics and Applications

In the sequel, we give an overview on some of these projects. The intention is to demonstrate in which areas RISC-Linz sees its expertise within robotics, to list at least some of our industrial cooperations, and to show the fundamental research topics that are investigated in this context. The intention is not to describe details of results here. For this purpose we refer, for each topic presented, to the most relevant publications written at the institute.

Gröbner bases for computing inverse kinematics: Gröbner bases are an algebraic method in polynomial ideal theory that provide a tool to solve a lot of problems concerned with the solution of algebraic systems of equations very elegantly. Gröbner bases can also be used to solve and analyse the inverse kinematics problems of open kinematic chains. The strategy for solving inverse kinematics problems by means of Gröbner bases is the following:

- The inverse kinematics problem is formulated as a system of equations based on Denavit-Hartenberg matrices.

- By either substituting new variables for sinus and cosinus or by using Cayley substitution, the system is reformulated as an algebraic system of equations.

- Gröbner bases can be used to triangularize this system of equations, to analyse the solution set of the system of equations, and to compute the solutions itself.

- Some of the parameters of the robot can be kept as variables during the computation, e.g. the lengths of links and/or the position of the endeffector. This allows to preprocess the solution for a whole class of robots.

Besides experiments with different versions of the Gröbner bases algorithm, special emphasis is put onto the analysis of the solution set (e.g. how can singularities in the motion be detected).

Relevant publications: (Stifter and Kutzler, 1988), (Stifter, 1992a).

Supervision system for tunnel constructions: In cooperation with the company Mayreder Consult, Linz, a software package for the simulation and automatic control of tunnel constructions has been developed. The tunneling method considered is a method of continuous excavation that involves only one type of ring for any course of the tunnel. The software environment supports the engineer in tunneling, i.e. allows the engineer to design the tunnel geometry, develop the course of the tunnel, the construction plans for the rings and ring elements, on the one hand, and to pursue the building process and the motions of the machines in the tunnel, and to control the machines, on the other hand. From a robotics point of view, especially the installation of the ring elements by a so called "erector" is of interest. This erector is a heavy load robot in its nature that has three base joints and three gripper joints. The joints can move all at the same time but their motions can, up to now, not be synchronized. One of the control problems for this robot is the path finding problem for picking up the ring elements from the conveyor belt, transporting them to the exact installation position and assembly them. We describe the idea for the solution of this challenging path finding problem in the next paragraph.

Relevant publications: (Schulter, Stifter, 1993), (Stifter, Neuwirth, 1990).

Path planning for non-synchronized motions: There is a great number of papers in the literature dealing with collision free path planning for robot manipulators. All these papers assume that the motions in the single joints of the robot are synchronized. This means that one assumes that, in case two or more joints are moving in parallel, one knows the positions of the joints relative to each other at each time moment. However, this must not be assumed for all robots, e.g. it is not achievable for the erector used for ring installations in tunneling. In case no synchronization is possible, but a simultaneous motion of two or more joints is desirable, the motions for the single joints have to be planned in such a way that collisions are impossible for all relative motions. This makes it necessary to stop motions at "synchronization points" in order to ensure a certain "structure" of the path. Typically, for non-synchronized motions, a path can only be specified by a sequence of intermediate positions, the "synchronization points". For the path between two such positions it is only guaranteed that the motion in each joint is monotonous (if a joint does move at all). Between two intermediate positions, there is no information about the relative positions of the joints. Hence, the goal of path planning for non-synchronized motions is to find intermediate positions such that any possible motion between them is collision-free. We call such paths "safe paths".

We treat the problem in configuration space. So we assume that we are already given a representation of free space and two positions, s and t, the start and goal position of the desired path. The goal is to compute a sequence of points in free space – the sequence of synchronization points – such that all componentwise monotonous curves between two consecutive synchronization points are contained in free space. We assume that free space is given as a (set of) polyhedrons with edges parallel to the coordinate axes. The algorithm works in the following steps:

Partition free space into isothetic rectilinear boxes (so called cells) by cutting free space by the hyperplanes that contain the faces of the polyhedrons representing free space. Clearly, if start and goal position of a path are in the same cell, these two positions specify a safe path. So finding a safe path can be reduced to finding a sequence of cells such that two consecutive cells have at least one point in common, which can be taken as synchronization point. A graph search technique can be used to find such a sequence of cells, taking the cells as vertices of the graph and connecting two cells that have a point in common by an edge. Different optimality criteria can also be taken into account by adjoining different weights to the edges. Especially keeping the number of synchronization points small (minimal) is important since at each synchronization point the whole motion has to be interrupted. This can be achieved by dynamically adjoining weights to the edges.

Relevant publications: (Stifter, 1992c), (Stifter, 1993).

Roider method for collision detection: The Roider method is an iterative algorithm that tests whether two objects (convex sets having inner points, that are characterized by the availability of certain "basic operations" on them) intersect. By practical experiences it turned out that the Roider method is quite efficient and applicable.

Roughly, the Roider method in 2D works in the following steps:

- Consider two objects A and B.
- Take a point P on the boundary of A.

- Construct the tangents from P to B, together with the touching points, T_1, T_2 on B and the additional intersection points, S_1, S_2 with A.

- If on both lines the sequence is $S_i P T_i$, then the objects are disjoint and a witness to disjointness has already been found.

- If on one of the lines the sequence is $S_i T_i P$ then T_i is contained in A and B, so the objects intersect.

- If none of the above situations is yet reached, then take S_i that is between P and T_i as new point P and repeat the procedure.

- One can prove that this iteration always stops, if the objects do not touch. In case they touch, the algorithm is stopped with the answer "too close together" as soon as the distance between P and some T_i is less than a prespecified tolerance.

Although the method is originally developed as an intersection test, in certain situations one can also ensure that a whole motion is collision free. These situations can be characterized by the relative positions of two objects in consecutive snapshots along the path and the direction, the length, or the course of the path. Since the statement for a whole path is based on a statement about two snapshots it is indispensable that there is some information available about the relative positions of the objects in the two snapshots. The knowledge that the two objects do not intersect would be not sufficient. In our case, the additional information is a wedge that contains one of the objects but nothing of the other, a "witness to disjointness".

Relevant publications: (Roider, Stifter, 1987), (Stifter 1988), (Stifter 1992b).

Robot vision based on geometric modeling: In cooperation with Joanneum Research, Graz, and the Technical University Vienna, we are working on a vision system for the recognition of technical objects. In the proposed system exisiting methods for 3D vision should be extended: on the one hand by the introduction of excessive feedback cycles in the chain "segmentation → matching → reconstruction → recognition" and, on the other hand, by the introduction of methods of the area of 3D modeling into the recognition part. The system will mainly be based on the following two starting points:

1. Active perception and active synthesis: The proposed system consists of a feedback loop. The initial recognition provides a depth interpretation and proposes an initial 3D model. This initial 3D model can be seen as a synthesis of an internal representation of the outside world. The synthesis is performed in an active manner, where the model is continuously updated and refined by the vision system which continuously searches for new information. The initial model again influences recognition, which provides a better depth interpretation.

2. Extraction of geometric models: In the construction of a geometric model from sensor data, no relevant information must be lost. Especially, if the sensors are mounted on moving parts, the geometry and kinematics of the

robots have to be taken into account. The research area of constructing geometric models for robots and their environments has been investigated in the last few years by many research groups. The particular choice of a modeling scheme as the internal representation of information from sensor data has to be done quite carefully. We base our system on the dexel modeling technique, an instance of spatial enumeration techniques that is especially close to visualization and to sensor information too.

Relevant publications: (Pölzleitner, Stifter, 1990), (Stifter, Taferner, 1992).

3 References

Pölzleitner, W., Stifter, S., (1990). Active sensing using associative processes and 3D geometric modeling; Proc. ÖAGM Tagung, Image Acquisation and real-time visualization, May 1990, Salzburg, Austria, Oldenbourg Verlag, pp. 139–151.

Roider, B., Stifter, S., (1987). Collision of convex objects; Proc. EUROCAL'87, Leipzig, GDR, June 1987, J. Davenport (ed.), Springer LNCS 378, pp. 258–259.

Schulter, A., Stifter, S., (1993). A simulation and demonstration package for tunnel construction; Journal of Civil Engineering. To appear.

Stifter, S., (1988). A generalization of the Roider method to solve the robot collision problem in 3D; Proc. ISSAC'88, Rome, Italy, July 1988, P. Gianni (ed.), Springer LNCS 358, pp. 332–343.

Stifter, S., (1992a). Computation and analysis of inverse kinematics by means of Gröbner bases; International Journal of Laboratory Robotics and Automation, Special Issue on Robot Kinematics, J. Lenarcic (ed.), vol. 4, pp. 115–125.

Stifter, S., (1992b). Collision detection in the robot simulation system SMART; Journal of Advanced Manufacturing Technology; vol. 7, pp. 277–284.

Stifter, S., (1992c). Path planning for non-synchrinized motions; Technical Report, RISC-Linz series no. 92-41.

Stifter, S., (1993). Shortest non-synchronized motions – parallel versions for shared memory CREW models; Proc. 2nd ACPC Conference, Gmunden, Austria, October 1993. To appear.

Stifter, S., Kutzler, B., (1988). Using symbolic methods in robotics: inverse kinematics and collision problems; Proc. IMACS Symp. on System Modeling and Simulation (SMS'88), Cetraro, Itally, Sept. 1988, S. Tzafestas, A. Eisinberg, L. Caotenuto (eds.), North Holland, pp. 229–234.

Stifter, S., Neuwirth, S., (1990). GET – a workbench for tunnel geometries; Proc. IASTED Int. Symp. Applied Informatics, Insbruck, Austria, Feb. 1990, Acta Press, h.M. Hamza (ed.), pp. 280–283.

Stifter, S., Taferner, S., (1992). The impact of geometric modeling on robot vision; Proc. ÖAGM Meeting on Image Acquisation and Real-Time Visualization, May 1992, Vienna, Austria, Oldenbourg Verlag, H. Bischof, W.G. Kropatsch (eds.), pp. 214–222.

Applied Robotics Group Research Activities at the ITIA-Institute for Industrial Technologies and Automation of CNR

C.R. Boër, E. Imperio
ITIA-Istituto di Tecnologie Industriali e Automazione
CNR-Consiglio Nazionale delle Ricerche
Milano, Italy

Abstract: This paper is a survey on the principal research activities carried out by the Applied Robotics Group of the Istituto di Tecnologie Industriali e Automazione (ITIA) of the National Research Council of Italy (CNR). ITIA activities include applied research, research training, innovation services, coordination and management of national and international projects concerning products, process technologies and production systems. By working in cooperation with companies, universities and research centres, the Applied Robotics Group is involved, in accordance with the Institute philosophy, in the frame of national and international projects such as Progetto Finalizzato Robotica, Brite-Euram, Eureka-Famos, China-CEE Cooperation, Central European Initiative. The aim is to contribute to the development and optimization of new components and sub-systems for the Automated Factory, both in manufacturing and assembly. This paper presents research themes with objectives, results, technological framing, partners and future outlook.

1. Introduction

Founded in 1923 as a non-profit body and reorganised in the 1945, the National Research Council (CNR) is an organ of the State which promotes, coordinates and regulates research in the interest of Italy's scientific and technological progress. To reach those goals, the CNR:
- coordinates national activities in the various branches of science and its application;
- set-up and fund scientific laboratories;
- fund and carry out research of national importance;
- assist scientific institutes, scholars and researchers through grants, scholarships and prizes;
- collect and publish scientific document and bibliographic material;
- in collaboration with Ministry of Foreign Affairs (MAE) and Ministry of Universities and Scientific and Technological Reasearch (MURST), handles Italy's participation in international scientific and technical organizations and programmes.

The institutional scientific activity of CNR is carried out directly by Operative or Research bodies: 192 Institutes (organisations of permanent nature, having as their aim direct research activity in relation to programmatic objectives of the CNR), 124 Study Centres (permanent bodies formed in the universities, scientific organisations, public administrations or private, for the development of particular studies and advanced research), 18 Research Groups (temporary organizations to carry out research leading to the organisation of the research work of many people and scientific bodies), 17 Finalized Project (quinquennial plan bodies for the achievement of objectives of great socio-economic interest, at a national level, through the involment of all the components of italian scientific network), 13 Strategic Projects (temporary initiatives mainly directed towards the capacity for reaserch within CNR itself on priority themes of national and international importance). The activity and running of the research bodies are controlled by 15 National Advisory Commitees covering the humanities and social sciences, as well as science, engineering, agriculture, medicine and the environment.

2. The ITIA: goals and activities

The Institute for Industrial Technologies and Automation (ITIA) of CNR, referring to the National Advisory Committee on "Technological Research and Innovation", was founded in 1963 and assumed, in 1979, its present status as an Institute. Starting from a long tradition of research on mechanical components for machine-tools, the Institute has devoted, in the recent years, its interest towards Factory Automation, increasingly based on automated design, automated production systems and innovative production technologies. The applied research activities are carried out, mainly, within the framework of national or, increasingly, international projects, such as Esprit, Brite and Eureka, in cooperation with Companies, Universities and Research Centres, both public or private. They covers the following scientifical and technological areas: CAD, Intelligent Planning Systems, Laser, Applied Robotics, Structural Analysis, Innovative Materials and Components, Reliability and Diagnostics, Technological Innovation. The Institute is directly involved in the coordination of the following national and international initiatives:

- the National Programme "Progetto Finalizzato Tecnologie Meccaniche"
- the National Programme "Progetto Finalizzato Robotica"
- EUREKA-FAMOS Umbrella Project (italian participation)
- EUREKA-MAINE Umbrella Project (italian participation)
- IMS Programme (italian participation).
- CHINA-CEE Cooperation (italian participation)

3. Applied Robotics Group of ITIA: research activities

The strategic approach of all research activities refers to study, design, realization and optimization of new hardware and/or software solutions for industrial automation, both for manufacturing and assembly applications. The philosophy that shares all research activities is to contribute to the development of the automated and flexible integration of the factory. The principal technological areas of research and development, in which the Applied Robotics Group operates, are: process planning, sensor systems and components for robot, assembly automation. Here below it will be presented the main results and research activities in progress.

3.1. Flexible Assembly Center (FAC)

A growing interest is shown in the automation of assembly processes, but the results have been particularly inconclusive for the assembly of complex and precise mechanical groups, in small-medium lot sizes, where the industrial robot has failed due to lack of precision and flexibility. The Group is studying a new concept for a Flexible Assembly Center to be integrated in a CIM environment, to obtain a full computer-automated, optimized, integrated and flexible assembly process. It is necessary to think to a new concept for an Assembly Center with characteristics similar to Manufacturing Center in terms of flexibility of management and control and rigidity in structural behaviour. Studies are under way to define more precisely the working environment of the Assembly Center and the design of a prototypal unit to show its feasibility. The FAC is based on the two main needs for the assemby of mechanical precise groups: flexibility and precision. The flexibility is needed for the assembly of different group and part configuration that is for the assembly of small lot size and even lot of one group. To achieve this condition it is necessary to have a set of standard and dedicated grippers readily available for the assembly center and furthermore the gripper change should be rapid and precise. It is also necessary to have a pallet changer for both the incoming parts and for the outgoing assembled group. The preparation and the setup of the parts should be done outside the machine in order to create parallelism during the assembly operation. The precision is dictated by the surface finish and tolerances of the mating parts. To achieve this condition it is necessary that the machine is rigid and the axis movements are precise that is within the tolerances required by assembled parts. The FAC is radically different form the known industrial robot technology: basically the configuration of the assembly center will be similar to a machining center (either vertical or orizzontal) with automatic modular pallet and/or fixtures changer and automatic tool(gripper) changer. Because the FAC, today, doesn't exists, it is an advantageous opportunity for the Applied Robotic Group to work on the possibility to introduce, more easily, some new tecnological concepts to design the Center, such as Group Technology for Assembly, Design for Assembly, Modularization of Fixtures, Numerical Control for Assembly and Concurrent Engineering.

3.2.   Automated Planning for Manufacturing Process

In the recent years, the growing push towards the factory automation and the related management of the information flow has marked the need to keep the company know-how which normally is own by one or few people with more years experiences and activity. The unfavourable loss of this knowledge, which may be caused by any reason, and the necessity of defining and making workable operational manufacturing standards, has induced the managers to adopt hardware and software tools more and more powerfull and intelligent to aid the design and the manufacturing. The objective of the work is to give a support to the automatic planning of the process necessary to manufacture mechanical parts. Chip removal processes have considered. In the overall process planning there are three aspects that are presently taken into consideration applying Artificial Intelligence techniques in order to reduce the level of complexity using traditional methods of programming. The first aspect is regarding the study of a CAD design for a mechanical component and the automatic extraction of the form features usefull for the production. In order to solve this problem a software module has been realized that recognizes the design features, it traslates them in a boundary representation and it extracts the manufacturing features. The second aspect is regarding the tool selection and the definition of the cutting parameters for the turning, drilling and milling operations. Also in this case the optimal definition of the problem is automatic and it is based on the knowledge base implemented with Expert System techniques. The third and last aspect regards the implementation of the working cycles taking into consideration the different factors with the importance is graded based on experience. Also in this case the automation of this phase is based on a knowledge base coded with rules and the use of research algorithms for the optimal solution. The originality of this work is based on the automation of each phase and the integration of every module in a complete software package. The program requires the use of a Personal Computer and the Windows operating system with a user interface easy to use. The research activity is made in cooperation with an industry leader in this domain in order to create the knowledge base and it is still in progress. A further development is considered to generalize the problem in terms of parts form and process type.

3.3.   Force/torque sensor application for robotized deburring

One of the most important scientific and technological field in which the Group is involved is, on one hand, the study, design and realization of force/torque sensors and, on the other hand, the study on application of similar sensor of the market. The project about sensors for robotized deburring, of which part of the research activitiy is carried out within a Central European Initiative Project (Working Group on Science and Technology/Committee on Industrial Technologies and Automation) in cooperation with the Academy of Science of Budapest and the Jozef Stefan Institute of Ljubljana, represents a good test for the integration of an industrial robot with a F/T sensor. The burr is variable, within a given interval, in form, dimension and position, and therefore it is a serious obstacle to the automation of the deburring process. A multiaxes force sensor is potentially the right step towards automation because flexible of easy reconfigurability. The actual studies are concentrated on the care of the problem: the interaction between the manipulator, the sensor, the tool and the burr.

3.4.   Eureka/Famos Project-EASY AIR:

Actually, the project is at feasibility study level and it is performed by italian-swedish consortium of partners. The extended title is: "Flexible and Integrated cell for Air Conditioner Subassembly Construction in a CIM environment" and the task of the above study is to reach a good organisation and integration of the activities to produce small heat-exchargers for ambient air conditioners that are the most characterizing perts of the machine. Inside teh coil fabrication flow, a process that hasn't been solved yet by any world company is the "hairpin insertion". The principal aim is to create an integrated cell able to perform automatically the hairpins insertion inside the fin pack.

3.5.   China-CEE Cooperation Program on CIM

The cooperation between China and CEE in the field of CIM started at the beginning of 1992 after almost one year of preparation; it will last another two years and sholud help to bring togheter two different ways to approach CIM. The cooperation program is a mixed structure of the China and CEE approaches. A consortium is formed at the european level including research institution from France (main contractor Country), Ireland, Germany, Belgium and Italy. The program is divided in four projects and each project is composed of an european and chinese institution working together on follow specific themes: CIM Demonstration, CIM Methodology, CIM Engineering Integration, CIM Architectureand Standardisation. The ITIA, through the Applied Robotics Group, partecipates to the first project, the final objective of which is to realize the handling-machining integration in a shop-floor environment, in CIMS-ERC of Tsinghua University in Beijing and to promote the use of simulation for design and control in industry. The global plane of this project includes work on Cell integration, workstation and cell communication systems, 3-D simulation and off-line programming and the development of a Manufacturing Process Simulator. The project has helped to bridge the cultural differences between the two regions and in terms of reasearch the first results are promising. The intention is to bring now some of the results in an industrial environment and the first steps in this direction are under way.

4. References

Boër, C.R., El-Chaar, J., Imperio, E., Rinaldi, R. (1990).Study for the assembly of Groups composed of very small parts. In: International FAMOS Workshop, Besancon, 145-162

Boër, C.R., El-Chaar, J., Imperio, E., Avai, A. (1991). Criteria for optimum layout design of assembly systems. In: Annals of CIRP, 2, 415-418

Boër, C.R. (1991). Industrial and academic experiences of CIM implementation. In: International Workshop on CIM-Strategy & Solution. Beijing

Boër, C.R., El-Chaar, J., Imperio, E. (1992). What is missing for a truly computer integrated assembly and manufacturing. In: International Symposium on Computer Integrated Manufacturing Systems-ISCIMS '92, Beijing

Boër, C.R., Imperio, E. (1992). Progettazione integrata: una soluzione per le aziende. In: Tecnologie Meccaniche, 6, 278-283

Boër, C.R., Imperio, E. (1992). La simulazione dei sistemi di produzione. In: Tecnologie Meccaniche, 7, 150-153

Boër, C.R., El-Chaar, J., Rinaldi, R., Vertova, L (1992). Un approccio all'utilizzo del sensore di forza per un sistema flessibile di sbavatura. In: ANIPLA Proceedings, 2, 550-562

Boër, C.R., El-Chaar, J., Galli, G., Imperio, E. (1992). Design and testing of a gripper for the automatic and flexible assembly of a family of precise mechanical components. In: Robotics and Manufacturing. Asme Press Series, 4, 553-558

Boër, C.R., Annacondia, E., Imperio, E. (1993). Verso la progettazione automatica. In: Tecnologie Meccaniche, 4, 162-165

Boër, C.R., Avai, A., Imperio, E. (1993). Experience of cooperation between China and CEE in the field of CIM. In: International Conference on Computer Integrated Manufacturing Systems-ICCIMS '93, Beijing

Robotic Research at the Scientific Academy of Lower Austria

P. Kopacek, G. Krenn
Department of Systems Engineering and Automation
Scientific Academy of Lower Austria, Krems, Austria

Abstract: One part of the Department of Systems Engineering and Automation at the Scientific Academy of Lower Austria is related to robotic science. There are three major groups working on different topics. These topics are: optimal path planning and control of industrial robots where special interest is laid on fuzzy controllers and neuronal networks, calculation and creation of flexible robotic assembly cells tasks based on the usage of databases and the creation of an expert system to handle errors in an assembly cell.

1 optimal control and path planning of a robot's motion

This project tries to find an optimal way to control robots.

To achieve this, many different topics in the control of the movement of robots are investigated. There are studies on the different ways in robot control and the planning of the robot's motion.

The different already known efforts are collected, compared and evaluated to find the most suitable way for controlling and planning the robot's movement. Deeper investigations are held on the part of fuzzy control and artificial neuronal networks, which lead to good results dealing with the nonlinearities of robotic systems.

As basic mathematical element the efficient and fast evaluation of the drive equations is examined.

Today there are many different ways in the control of the path- following of the robot.

One group of concepts deals with algorithms to evaluate the necessary control signals from a mathematical model of the robot.

Another Group of control systems uses more sophisticated models and provides a feed forward compensation. In a first step couplings between the links are neglected and an optimal controller is designed. In a second step an additional moment M, evaluated from the drive equations is added to the first control signals and compensates the internal couplings between the links.

The most sophisticated group of this kind of controls uses terms of non-linear decoupling or inverse systems. It is also based on model equations of the system but the control signals are signed to the system in a way that it behaves like separate linear decoupled systems of similar dynamic behaviour.

A special group of control designs deals with adaptive controls.

There are different subgroups in the field of adaptive controls:
- methods using a reference model
- self tuning algorithms
- model based, predictive control

The group of methods using a reference model can further be split into
- control concepts that adapt the reference model (these are the slower ones)
- MRAC (Model Reference Adaptive Control)

The MRAC approach adapts the control in a way that robot + control behave like the ideal model, only the controller's parameters are changed continuously.

The methods of predictive control differ in some specialities, but the basic elements are quite the same. The system consists of a model based controller which has been extent by a predictor to compensate the modelling and discretization errors .

A quite new part of control technology deals with fuzzy methods and artificial neuronal networks.

Fuzzy methods are used in highly complex, nonlinear, hard to model systems like an industrial robots.

The advantage with the use of fuzzy control structures in robotics lies in a better performance with nonlinear processes. The fuzzy controller is used as common nonlinear discrete controller. It consists of some basic modules depicted in Fig.1.

In standard fuzzy control loops the control deviation is evaluated and converted into fuzzy compatible form. Then the control signal is built by using fuzzy rules of reasoning. After the decision the fuzzy control signal is computed into a deterministic control signal output.

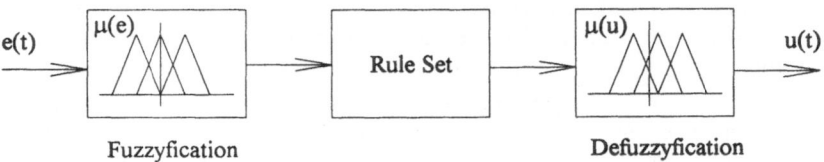

Figure 1: Structure of a Fuzzy Controller

The conversion into and back from fuzzy form is done by the means of so called membership functions, which have to be partially linear.

The main problem with the design of fuzzy concepts for robots is the optimisation of the fuzzy controller. In most cases this means a heuristic tuning of the fuzzy controllers by "Trial and Error".

By using the methods of nonlinear control theory a way to find a possibility for a real optimisation of fuzzy controllers and the exact formulation of the optimisation task is being searched. The optimisation of only one setpoint change has already multicriterical character. The multicriterical problem is transformed to a siglecriterical one by using the method of the weight sum of paricular criteria.

With the tool ILMFUZZY it is possible to create and optimise a fuzzy controller for typical nonlinear systems. The results show that the fuzzy controller is superior to the PID controller.

Another quite new way of controlling robots is the use of artificial neuronal networks. These elements can be used in control circuits in various forms:

The usage of ANNs in model based robotic control helps to eliminate some problems which occur by the implementation of conventional controllers.

There are problems like

- The need to make simplifications in respect to the complex calculations of a dynamical model by using a closed loop control circuit

- A model that reproduces the whole dynamical behaviour of a robotic system does not exist (e. g. friction, loose, hysteresis)

- Changes in the dynamical behaviour caused by wear

Because of their special structure ANNs have some outstanding characteristics which make them well suited for solving the problems mentioned above:

- The parallel structure of ANNs allows their implementation in complex real time systems

- The ANN's ability to learn allows adaptation to the changing dynamical behaviour of a robot

- The ability to recognise the important part of an information allows ANNs to make the robot control unsensitive against disturbances and measuring errors.

Ways to insert The ANN into control circuits are:

- Replacement of a conventional controller by an ANN

- Usage of the ANN in feed forward circuits: the ANN influences the robot in a way to improve accuracy by covering ranges which are badly covered by the linear controller.

- ANN in model based controls: the ANN is used to evaluate the nonlinear dynamical connections inside the model based control algorithm.

Besides the investigations on control concepts the different methods of fine motion planning are analysed.

Optimal planning and execution of the robot's motion is a very important field of robot control because great amounts of time can be gained or lost depending on the algorithm that is used.

Starting with Paul's algorithm for Cartesian path control different advanced and higher sophisticated fine motion planning methods are examined.

2 Software tool for the semiautomatic design of small assembly cells

A second group of the department deals with the calculation and creation of flexible robotic assembly cells tasks of one or two robots.

An optimal, timesaving work on offers reduces the non-productive time. This reduction is very important especially during times of economic recessions. The design of small assembly cells with a maximum of two robots is a typical recess for small and medium sized companies. They have to react fast and flexible to win most of the requested offers.

The layout of the assembly cell is being designed with the semiautomatic and database supported Planning Software. It is possible to compute the cyclic time and the price of the assembly cell as well as the maximum delivery time of the components. Additionally, it is possible to generate an export file for the AutoCAD-based simulation software SITAR.

Graphical User Interfaces like MS-Windows (Version 3.1) are used to plan the applications. The Planning Software requires a relational database to store the total input and output information that is necessary for the planning progress. SQLWindows as a MS-Windows database frontend development tool in connection with the SQL database backend of ORACLE (Version 6.0) is used by the PC-GUI-based Planning Software.

Four relational connected database parts in three different layers of hierarchy are forming the support database. These four main parts of the database are:

- Planning Database: requires data from all other databases. Each result generated by the interactive Planning Software is written into the Planning Database

- Product Database: stores the general product, manufacturer, reference person and branch data plus additional product information

- Symbol Database: the Planning Software is icon driven and each necessary icon data is stored here

- Component Database (21 tables): Each component which is used from the Planning Software is stored in this treelike database. As the largest of the four database parts it includes general and specific data of a component.

The process of generating and calculating the process parted into subgroups like the description of the product and input of target values, the description of the assembly-manufacturing operations and definition of their order, the splitting of assembly - manufacturing operations between the robots and the selection of tools and grippers, refinement of assembly - manufacturing operations and coarse cyclic time and the input of required components.

As result of the planning process the price, cyclic time and delivery time is calculated and can be compared with the target values. If the results are sufficient, the user can accept the results and create a printout of required components or drop the results and restart the planning process. A list of all selected components could be printed on the screen and the printer.

3 An expert system for diagnosis in assembly cells

A third group of colleagues works on expert systems.

As result of their work an expert system for diagnosis in assembly cells has been created.

The system allows less skilled workers to set the required actions in order to solve errors in the assembly cell. In addition to the assistance of supervising persons in solving plant errors, the knowledge based system can also be used for tutoring workers by simulation of various conditions of the assembly plant. The knowledge base is completed and maintained by the person who looks after the plant. It is possible to adapt the diagnostic routine to extensions in the assembly cell.

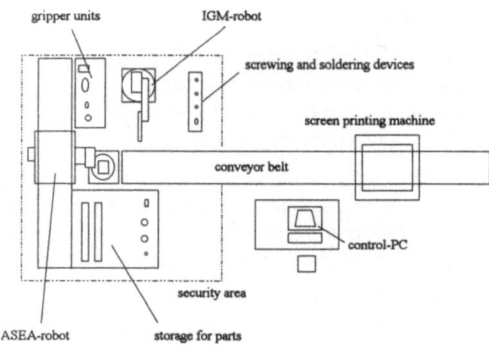

Figure 2: assembly cell

The knowledge based system is running on the control - PC of the assembly system. It is available in any case of error.

The principal item of the plant are two robots for assembly, which can be armed with various tools like screwing units, robot grippers and a soldering device. The cell further includes an automatic tool-changing-system, storage devices, helical conveyor units, a conveyor belt and a screen printing machine as shown in fig. 2.

The whole system is controlled by a control software running on the PC. In case of machine failure the error detection module can be called by a function key. Both robots stop in their momentous position and after the control programme is stopped, all actual parameters are saved and the error diagnosis programme is started. During the diagnosis session the errors are identified and corrected by the user in interaction with the system. In a further step the worker has to eliminate the error reason by selecting suitable actions out of the menus and by manual intervention (e. g. removing defective parts...)

For this feature an interface between the control programme and the external diagnosis programme had to be created. To make possible an error detection the state arrays of the programmable controller, which describe the actual state of the whole assembly cell, are evaluated by the control programme, written to protocol files and passed to the error detection.

Experts' knowledge and experience has been collected by interviews and by accompanying forms. Additional knowledge has been extracted from existing error-protocols. This knowledge has been being transformed into logical rules. Working with the system is quite easy. In case of an occuring error the machine starts to test various fields of the state arrays which leads to a coarse classification of the error. Further decisions are made in interaction with the user. The programme asks the user to check different parts of the assembly cell. The questions are simple and the user has to answer by typing the Yes or No key. Usually not more than 10 questions have to be answered until a solution is found. If the solution of the detected problem is not known to the user he can ask for repair hints.

References

Troch, I. (1992). Robot Control. Summer School 1992 at the Scientific Academy of Lower Austria

Desoyer, K. (1992). Geometrie, Kinematik und Kinetik von Industrierobotern. Summer School 1992 at the Scientific Academy of Lower Austria

Kopacek P., Troch, I., Desoyer, K. (1988). Theory of Robots. IFAC Proceedings Series Number 3

Paul, R. (1979). Manipulator Cartesian Path Control. IEEE Trans., SMC- 9, 702- 711

Lee, C., Gonzalez, R., Fu, K. (1983). Tutorial on Robotics. IEEE Computer Society Press/North-Holland

Koch, M., Kuhn, Th., Wernstedt, J. (1993). Eine neue Entwurfmethode für Fuzzy Regelungen. at 41/5, 152-158

Kosko, B. (1992). Neural Networks and Fuzzy Systems. Prentice Hall, Inc.

Kuhn, T., Wernstedt, J. (1992): ILMFUZZY Version 1.0 Ein Programmpaket zur Simulation und Optimierung einschleifiger Fuzzy Regelkreise, TU Ilmenau 1992

Tzafestas, S.G. (1991). Intelligent Robotic Systems. Marcel Dekker, Inc., New York, 1991

Beneder, M. (1992). An Expert System for Diagnosis in Assembly of Welding Transformers (in German), Scientific Works Series, Vol. 3, SAT, Krems.

Robot Wrist Configurations, Mechanisms and Kinematics

A. Romiti, T. Raparelli, M. Sorli[2]
Department of Mechanics
Politecnico di Torino, Italy
[2] Department of Industrial Engineering
University of Cassino (FR), Italy

Abstract: This paper discusses a general method for determining the kinematic performance of spherical robot wrists. Different wrist types considered include R-P-R (roll-pitch-roll) and P-Y-R (pitch-yaw-roll) wrists. Singularity conditions are indicated for both cases. For R-P-R wrists, singularity is defined by the ratio of the angular velocity of each motor to the velocity around the degeneracy axis. For P-Y-R wrists, singularity is identified both by analyzing the Jacobian matrix and by analyzing the relative velocities of the wrist components.

1 Introduction

Robot wrists are designed to provide orientation to the end effector. They should preferably be centered around a point, that means to constitute "spherical joints", because the resulting configuration is more dexterous and less cumbersome than the other configurations.

Robot wrists should have a low "degeneracy level", that represents the region where some rotations around certain fixed axes in the cartesian space are forbidden, or require very high speeds of the actuators. Degeneracy occurs when the three axes of the spherical wrist joints are contained in one plane. The degeneracy condition implies that it is not possible to follow the shortest path in producing certain orientations according to a specified sequence in the work envelope without exceeding the maximum possible velocity for certain joints [Paul and Stevenson (1983)], [Huang and Milenkovic (1987)], [Treveljian et al. (1986)].

They are two main principles concerning the make of wrists. One is the R-P-R (roll-pitch-roll) method; the other is the P-Y-R (pitch-yaw-roll) method [Rivin (1988)]. R-P-R wrists are based on the application, through conical gears, of the Euler formulas for successive rotations of bodies with respect to the others [Litvin and Zhang (1986)], [Romiti and Sorli (1992)]. P-Y-R wrists act in different way. Two axes (P and Y) are fixed in space (with respect to the last robot arm): the adjiustment of rotations to comply to kinematic constraints is obtained by free rotations of bodies one with respect to the others [Milenkovic (1987)], [Rosheim (1989)], [Romiti and Raparelli (1993)].

2 R-P-R wrists

R-P-R wrists consist of three arms connected in sequence via three rotary joints. The kinematic connection of each arm with the preceding arm can be described using the transfer matrices $^{i-1}A_i$; if α_{i-1} is the angle which brings joint axis z_{i-1} to z_i through a rotation around axis x_{i-1}, and if θ_i is the angle which brings axis x_{i-1} to x_i through a rotation around axis z_i we have:

$$^{i-1}A_i = \text{Rot}(x_{i-1}, \alpha_{i-1})\text{Rot}(z_i, \theta_i) = \begin{vmatrix} \cos\theta_i & -\sin\theta_i & 0 \\ \sin\theta_i\cos\alpha_{i-1} & \cos\theta_i\cos\alpha_{i-1} & -\sin\alpha_{i-1} \\ \sin\theta_i\sin\alpha_{i-1} & \cos\theta_i\sin\alpha_{i-1} & \cos\alpha_{i-1} \end{vmatrix} \tag{1}$$

Angular velocities of the successive arms are given by the expression:

$$\vec{\omega}_i = \vec{\omega}_{i-1} + \vec{\omega}_{(i-1,i)} \tag{2}$$

where $\vec{\omega}_{(i-1,i)}$ describes the angular velocity of body i relative to body $i-1$. Assuming that the three sets are arranged so that the three degrees of freedom θ_i occur around axes z_i, we have:

$$\vec{\omega} = \dot{\theta}_1\vec{k}_1 + \dot{\theta}_2\vec{k}_2 + \dot{\theta}_3\vec{k}_3 \tag{3}$$

Remembering (1), we then have (where $C_i=\cos(\theta_i)$ and $S_i=\sin(\theta_i)$ with i=1,2):

$$\vec{\omega} = \begin{vmatrix} \omega_x \\ \omega_y \\ \omega_z \end{vmatrix} = \begin{vmatrix} 0 & -S_1 & C_1S_2 \\ 0 & C_1 & S_1S_2 \\ 1 & 0 & C_2 \end{vmatrix} \begin{vmatrix} \dot{\theta}_1 \\ \dot{\theta}_2 \\ \dot{\theta}_3 \end{vmatrix} \tag{4}$$

We will examine a spherical wrist developed at our laboratory. It has three axes which intersect at right angles (Figure 1a). Roll movement of arm 1 is controlled by the motor with angular velocity $\dot{\phi}_1$. Arm 2 is rotated around axis z_2 by the bevel gear set consisting of gears N_1 (connected to the motor with angular velocity $\dot{\phi}_2$) and N_2 (connected to arm 2). The roll movement takes place around axis Z_3 and is produced via bevel gears N_3, N_4, N_5 and N_6.
Indicating the axis perpendicular to z_2 and z_3 with x' and using the notation shown in figure 1b which defines the configuration assumed by the three sets of cartesian axes, we can write:

$$\begin{aligned} \omega_{x'} &= C_1\omega_x + S_1\omega_y \\ \omega_{y'} &= C_1\omega_y - S_1\omega_x \\ \omega_{z'} &= \omega_z \end{aligned} \qquad (5)$$

As the wrist has a single center, we have:

$$\frac{N_2}{N_1} = \frac{N_4}{N_3} = \frac{N_5}{N_6} \qquad (6)$$

$$\begin{aligned} \dot{\phi}_2 - \dot{\phi}_1 &= \frac{N_2}{N_1}\dot{\theta}_2 \\ \dot{\phi}_3 - \dot{\phi}_1 &= \frac{N_4}{N_3}\dot{\theta}_4 \\ \dot{\phi}_1 &= \dot{\theta}_1 \end{aligned} \qquad (7)$$

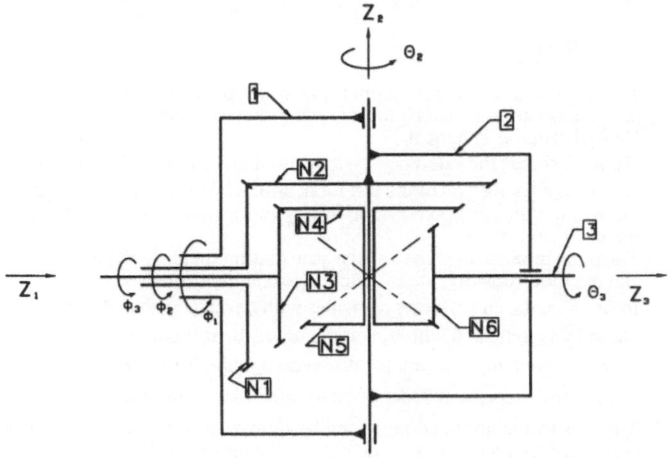

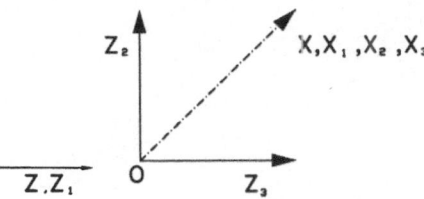

Fig. 1a. Kinematic scheme of a R-P-R wrist

The common angular speed of wheels N_4 and N_5 around axis z_2 is $\dot{\theta}_4$. The angular speed of wheel N_3, $\dot{\theta}_3$, is then given by:

$$\dot{\theta}_4 - \dot{\theta}_2 = \frac{N_6}{N_5}\dot{\theta}_3 \qquad (8)$$

The relations between velocities $\dot{\theta}_i$ and $\dot{\phi}_i$ with i = 1,2,3 are thus:

Fig. 1b. Wrist reference systems

$$\begin{aligned} \dot{\theta}_1 &= \dot{\phi}_1 \\ \dot{\theta}_2 &= \frac{N_1}{N_2}(\dot{\phi}_2 - \dot{\phi}_1) \\ \dot{\theta}_3 &= \frac{N_5}{N_6}\left[\frac{N_3}{N_4}(\dot{\phi}_3 - \dot{\phi}_1) - \frac{N_1}{N_2}(\dot{\phi}_2 - \dot{\phi}_1)\right] \end{aligned} \qquad (9)$$

Introducing equations (5) in (7) gives:

$$\begin{aligned} \dot{\phi}_1 &= \omega_z - \frac{C_2}{S_2}\omega_{x'} \\ \dot{\phi}_2 &= \omega_z - \frac{C_2}{S_2}\omega_{x'} + \frac{N_2}{N_1}\omega_{y'} \\ \dot{\phi}_3 &= \omega_z + \frac{1}{S_2}\left(\frac{N_4 N_6}{N_3 N_5} - C_2\right)\omega_{x'} + \frac{N_4}{N_3}\omega_{y'} \end{aligned} \qquad (10)$$

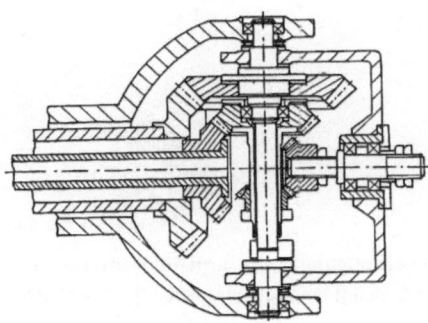

Fig. 2. R-P-R wrist constructive example

Note that possible degeneracy conditions are shown by the fact that it is not possible to rotate around axis x' for $\theta_2=0$. In other words, with θ_2 tending towards zero, the motors must have angular velocities ϕ_i tending towards infinity in order to achieve a finite velocity $\omega_{x'}$.

The ratios $\phi_i/\omega_{x'}$ define the degeneracy level of this R-P-R wrist as a function of angle θ_2.

A constructive solution for a wrist R-P-R, whose kinematic behaviour corresponds at the scheme of figure 1a, is shown in figure 2.

Similar considerations should be made for other types of R-P-R wrist as discussed by Romiti and Sorli (1992).

3 P-Y-R wrists

The wrist analyzed here is a pitch-jaw-roll type. Unlike the majority of wrists of this type however, all movements are actuated by motors which can be mounted on the robot base and by belt and chain drives. Configuration is spherical.

Figure 3 shows the reference systems used and the wrist's kinematic layout: x,y,z is a set of Cartesian axes located at the end of the last robot arm, while $\vec{i},\vec{j},\vec{k}$ are the respective versors. Axis x coincides with the arm longitudinal axis, while axes y and z are transverse to the arm. The origin of the set of axes is the wrist center.

The pitch motion (in plane xz) and yaw motion (in plane xy) are transmitted to the members carrying the end effector, to which the roll motion is communicated.

In figure 3, the end effector coincides with bevel gear shaft 4, which is associated with versor $\vec{v}$; rotation around $\vec{v}$ is indicated with θ_3 and identifies the roll motion.

Yaw and pitch motions are identified by $\theta_2\vec{k}$ and $\theta_1\vec{j}$ respectively.

The axis of gears 2 and 3 identified by versor $\vec{\lambda}$ is connected by body 8 via rotating fits to the end effector.

The axis of the articulation carried by toothed quadrant 6 is identified by versor $\vec{\mu}$. This articulation is connected by body 7 to the end effector again through a rotating fit. This permits a relative adjustment motion between members.

Axis $\vec{v}$ is thus always normal to versors $\vec{\lambda}$ and $\vec{\mu}$. Versors $\vec{\lambda}$ and $\vec{\mu}$ move in planes xz and xy respectively.

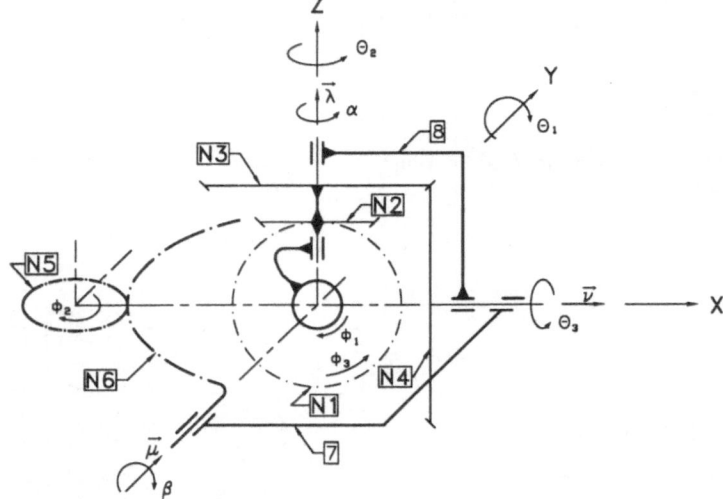

Fig.3. Kinematic scheme of a P-Y-R wrist

In a configuration obtained following a pitch and yaw motion, $\dot{\alpha}$ is the angular velocity of body 8 relative to $\vec{\lambda}$, and $\dot{\beta}$ is the angular velocity of 7 relative to $\vec{\mu}$. Thus:

$$\vec{v} = \vec{\mu} \wedge \vec{\lambda} = \frac{C_2C_1\vec{i} + S_2C_1\vec{j} - S_1C_2\vec{k}}{\sqrt{C_1^2 + S_1^2C_2^2}} \tag{11}$$

given that:

$$\vec{\mu} = -S_2\vec{i} + C_2\vec{j} \qquad\qquad \vec{\lambda} = S_1\vec{i} + C_1\vec{k} \tag{12),(13}$$

Angular velocities of bodies 7 and 8 are:

$$\vec{\omega}_7 = \dot{\theta}_2 \vec{k} + \beta \vec{\mu} \qquad\qquad \vec{\omega}_8 = \dot{\theta}_1 \vec{j} + \dot{\alpha} \vec{\lambda} \qquad (14),(15)$$

The angular velocity of body 8 relative to body 7 must be directed along $\vec{v}$ because of the constraint of rotary joints; we have:

$$\vec{\omega}_{8/7} = \vec{\omega}_8 - \vec{\omega}_7 = (\dot{\alpha}S_1 + \beta S_2)\vec{i} + (\dot{\theta}_1 - \beta C_2)\vec{j} + (\dot{\alpha}C_1 - \dot{\theta}_2)\vec{k} = \omega_{8/7}\vec{v} \qquad (16)$$

and then one has:

$$\dot{\alpha}S_1 + \beta S_2 = \omega_{8/7} \frac{C_2 C_1}{\sqrt{C_1^2 + S_1^2 C_2^2}}$$

$$\dot{\theta}_1 - \beta C_2 = \omega_{8/7} \frac{S_2 C_1}{\sqrt{C_1^2 + S_1^2 C_2^2}} \qquad (17)$$

$$\dot{\alpha}C_1 - \dot{\theta}_2 = -\omega_{8/7} \frac{S_1 C_2}{\sqrt{C_1^2 + S_1^2 C_2^2}}$$

which give the following two expressions:

$$\dot{\alpha} = \frac{1}{C_1}\left(\dot{\theta}_2 - \omega_{8/7} \frac{S_1 C_2}{\sqrt{C_1^2 + S_1^2 C_2^2}} \right) \qquad \beta = \frac{1}{C_2}\left(\dot{\theta}_1 - \omega_{8/7} \frac{S_2 C_1}{\sqrt{C_1^2 + S_1^2 C_2^2}} \right) \qquad (18),(19)$$

By stating values for $\omega_{8/7}$ in (18) and (19), we can obtain $\dot{\alpha}$ and β expressed only as a function of $\dot{\theta}_1$ and $\dot{\theta}_2$, i.e.:

$$\dot{\alpha} = \dot{\alpha}(\dot{\theta}_1, \dot{\theta}_2) \qquad\qquad \beta = \beta(\dot{\theta}_1, \dot{\theta}_2) \qquad (20),(21)$$

The absolute angular velocity of the end effector can be written in either of the following ways ($\dot{\theta}_3$' and $\dot{\theta}_3$" differ only by $\omega_{8/7}$):

$$\vec{\omega} = \dot{\theta}_1 \vec{j} + \dot{\alpha} \vec{\lambda} + \dot{\theta}_3. \vec{v} \qquad (22)$$

$$\vec{\omega} = \dot{\theta}_2 \vec{k} + \beta \vec{\mu} + \dot{\theta}_3.. \vec{v} \qquad (23)$$

In matrix form we have:

$$\vec{\omega} = \begin{vmatrix} \omega_x \\ \omega_y \\ \omega_z \end{vmatrix} = \begin{vmatrix} 0 & S_1 & \dfrac{C_1 C_2}{\sqrt{C_1^2 + S_1^2 C_2^2}} \\ 1 & 0 & \dfrac{S_2 C_1}{\sqrt{C_1^2 + S_1^2 C_2^2}} \\ 0 & C_1 & \dfrac{-S_1 C_2}{\sqrt{C_1^2 + S_1^2 C_2^2}} \end{vmatrix} \begin{vmatrix} \dot{\theta}_1 \\ \dot{\alpha} \\ \dot{\theta}_3 ' \end{vmatrix} = [J'] \begin{vmatrix} \dot{\theta}_1 \\ \dot{\alpha} \\ \dot{\theta}_3 ' \end{vmatrix} \qquad (22')$$

$$\vec{\omega} = \begin{vmatrix} \omega_x \\ \omega_y \\ \omega_z \end{vmatrix} = \begin{vmatrix} 0 & -S_2 & \dfrac{C_1 C_2}{\sqrt{C_1^2 + S_1^2 C_2^2}} \\ 0 & C_2 & \dfrac{S_2 C_1}{\sqrt{C_1^2 + S_1^2 C_2^2}} \\ 1 & 0 & \dfrac{-S_1 C_2}{\sqrt{C_1^2 + S_1^2 C_2^2}} \end{vmatrix} \begin{vmatrix} \dot{\theta}_2 \\ \beta \\ \dot{\theta}_3 '' \end{vmatrix} = [J''] \begin{vmatrix} \dot{\theta}_2 \\ \beta \\ \dot{\theta}_3 '' \end{vmatrix} \qquad (23')$$

The values of angles θ_1 and θ_2 at which the determinant of the Jacobian matrix is zero identify the singularity configurations. There are thus two possible constraint configurations, one for the first mechanism and one for the second mechanism. For the first mechanism, we have $\det[J']=0$ for $C_2=0$, i.e. for $\theta_2=\pi/2$. Angular velocity β cannot be given: axis $\vec{v}$ cannot rotate around x. For the second mechanism, $\det[J'']=0$ for $C_1=0$, i.e for $\theta_1=\pi/2$. Rotation $\dot{\alpha}$ is impeded: axis $\vec{v}$ cannot rotate around axis x. In both cases, the singularities could have been determined directly from (18) and (19), which show that β tends to zero for θ_2 tending to $\pi/2$, and that $\dot{\alpha}$ tends to zero for θ_1 tending to $\pi/2$. In any case, the axis of the degeneracy cones is perpendicular to the wrist's main longitudinal axis (directed along x), so that the wrist shows no degeneracy along the axis of the spherical surface which defines the size of the work space.

48

The relations between $\dot{\theta}_i$ and $\dot{\phi}_i$ with i = 1,2,3 are thus:

$$\dot{\theta}_1 = \dot{\phi}_1 \qquad \dot{\theta}_2 = \frac{N_5}{N_6}\dot{\phi}_2 \qquad \dot{\theta}_3 = \frac{N_1 N_3}{N_2 N_4}\dot{\phi}_1 + \frac{N_5 N_3}{N_6 N_4}\dot{\phi}_2 + \frac{N_1 N_3}{N_2 N_4}\dot{\phi}_3 \tag{24}$$

where the number of teeth on the i-nth gear is indicated with N_i. By introducing equations (20) and (21) in (22) or (23) and allowing for (24), it is possible to obtain a procedural analysis similar to that conducted for R-P-R wrists and thus indicate degeneracy levels. Figure 4 shows the constructive solution of the new P-Y-R wrist.

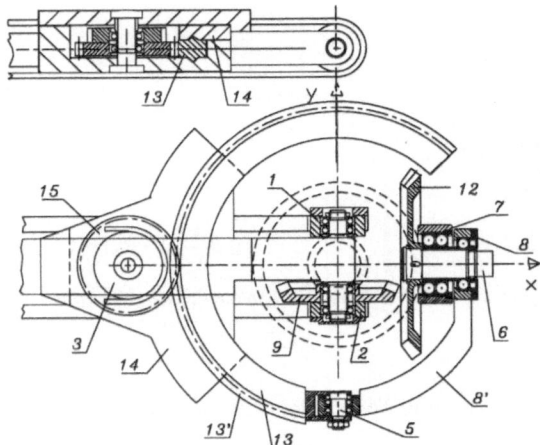

Fig.4. P-Y-R wrist constructive example

4 Conclusions

The kinematic performance of two different types of orthogonal axis spherical wrists with three degrees of freedom are analyzed and compared. The configurations of the wrists presented in the paper are original, but the analysis method can be readily extended to other types of R-P-R and P-Y-R wrists. Wrist degeneracy conditions are discussed, with particular reference to the position of the degeneracy cone and degeneracy level. The kinematic relations are then found which link the angular velocities of the motors actuating the three degrees of freedom to the component of the end effector absolute velocity in the set of three axes located on the terminal robot arm.

In this way, we think to offer an overview of wrist configurations and possibilities, and an instrument both for designing wrists and for forecasting and comparing their performances.

This research project was supported by National Research Council C.N.R. of Italy as part of the "Progetto Finalizzato Robotica".

Bibliography

Huang,B., Milenkovic,V. (1987). On an algorithm negotiating wrist singularities. Proc. of 17th Int. Symposium on Industrial Robots, pp.13-1,13-6.

Litvin,F.L.,Zhang Yi.(1986). Robotic bevel gear differential train. International Journal of Robotic Research, vol.5, n.2, pp.75-81.

Milenkovic,V.(1987). New non-singular robot wrist design. Proc. of 17th Int. Symposium on Industrial Robots, pp.13-29,13-42.

Paul,R.P., Stevenson,C.N.(1983). Kinematics of robot wrists. International Journal of Robotic Research, vol.2, n.1, pp.31-38.

Rivin,E.I.(1988). Mechanical design of robots. McGraw-Hill, pp.285-303.

Romiti,A.,Sorli, M.(1992). Wrist analysis and degeneracy evaluation and comparison. Proc.of 2th Int. Symp. on Measurement and Control in Robotics, Tsukuba, Japan, pp.437-444.

Romiti,A.,Raparelli,T.(1993). A new spherical robot wrist with remote degeneration regions. Submitted to 6th ICAR, Tokyo, Japan.

Rosheim,M.E.(1989). Robot wrist actuators. Wiley.

Treveljian,J.P.,Kovesi, P.D.,Ong, M.,Elford, D.(1986). ET: a wrist mechanism without singular positions. International Journal of Robotic Research, vol.4, n.4, pp.71-85.

Parallel Evaluation of Robot Kinematic Transformations

K. Dobrovodsky, P. Kurdel
Institute of Control Theory and Robotics
Slovak Academy of Sciences
Bratislava, Slovakia

Abstract: Real-Time evaluation of kinematic transformations consumes a significant amount of the processor time in a robot control system. The aim of the paper is to explain the possibilities of the parallel evaluation of a general direct kinematics and the principle of the parallel evaluation of an inverse kinematics. To complete the overview of the problem, some important benefits and drawbacks of the parallel implementation of kinematic transformations are presented.

1 Introduction

Many various formalisms have appeared to describe kinematics of robot manipulators. The matrix method is very popular and it is in common use. A symbolic notation permitting any position representation will be considered in the following text. Any open kinematic chain is uniquely defined by a sequence of constant and variable translations and/or rotations. Let's introduce the following abbreviations:

A, B, C	constant rotations	a, b, c	constant translations
X, Y, Z	variable rotations	x, y, z	variable translations

The above operators may be represented by $4 \cdot 4$ homogeneous transformation matrices $I(\delta), J(\delta), K(\delta), i(\delta), j(\delta), k(\delta)$ describing the relative rotation about or translation along one of the corresponding coordinate axes i,j,k, so that e.g.

$$I(\delta) = \begin{pmatrix} 1, & 0, & 0, & 0 \\ 0, & \cos(\delta), & -\sin(\delta), & 0 \\ 0, & \sin(\delta), & \cos(\delta), & 0 \\ 0, & 0, & 0, & 1 \end{pmatrix}, \qquad j(\delta) = \begin{pmatrix} 1, & 0, & 0, & 0 \\ 0, & 1, & 0, & \delta \\ 0, & 0, & 1, & 0 \\ 0, & 0, & 0, & 1 \end{pmatrix}$$

where δ is either constant or variable. Other position representations (quaternions, ternions) may be considered, too. We will not deal with possible parallel evaluation of sin, cos, sqrt and atan2 functions. Our goal here is to point out parallel possibilities coming from the robotic open kinematic structure.

Consider a sequence of small and capital letters $cZcAcXcXcZcaXcZcA$. This sequence defines the kinematic structure of an open kinematic chain with seven links: c, cAc, c, c, ca, c, cA which are connected by six joints: Z, X, X, Z, X, Z. An implicit indexing of constants $\lambda = (\lambda_1, \lambda_2, \ldots)$ and variables $\vartheta = (\vartheta_1, \vartheta_2, \ldots)$ is assumed from left to right. Thus the string $\Omega = cZcAcXcXcZcaXcZcA$ uniquely defines a parametric operator

$$\begin{aligned}
\Omega(\vartheta, \lambda) &= c(\lambda_1) \cdot Z(\vartheta_1) \cdot c(\lambda_2) \cdot A(\lambda_3) \cdot c(\lambda_4) \cdot X(\vartheta_2) \cdot \\
&\quad \cdot c(\lambda_5) \cdot X(\vartheta_3) \cdot c(\lambda_6) \cdot Z(\vartheta_4) \cdot c(\lambda_7) \cdot a(\lambda_8) \cdot \\
&\quad \cdot X(\vartheta_5) \cdot c(\lambda_9) \cdot Z(\vartheta_6) \cdot c(\lambda_{10}) \cdot A(\lambda_{11})
\end{aligned} \tag{1}$$

where $\lambda = (\lambda_1, \ldots, \lambda_{11})$ are constant parameters and $\vartheta = (\vartheta_1, \ldots, \vartheta_6)$ are variable parameters. The direct kinematics lies in specifying $\Omega(\vartheta, \lambda)$, given the parameters λ and the joint variables ϑ. The inverse kinematics lies in obtaining a solution for the joint variables, given the parameters λ and the homogeneous transformation $\Omega(\vartheta, \lambda)$.

2 Direct Kinematics

During the positioning of robot kinematic structure the robot control system is generating positions of servo drives, i.e. positions of joints. The direct kinematics transformation yields the cartesian coordinates of the position and orientation of the gripper affixed to the last link of the open kinematic structure. Evaluation of the direct kinematics consists of compositions of constant and variable transformations. Adjacent constant transformations (with λ_i) may be evaluated in advance ("off line"). According to the above example:

$$T_1 = c(\lambda_1) \qquad T_2 = c(\lambda_2) \cdot A(\lambda_3) \cdot c(\lambda_4) \qquad T_3 = c(\lambda_5)$$
$$T_4 = c(\lambda_6) \qquad T_5 = c(\lambda_7) \cdot a(\lambda_8) \qquad T_6 = c(\lambda_9)$$
$$T_7 = c(\lambda_{10}) \cdot A(\lambda_{11})$$

Variable terms $Z(\vartheta_1)$, $X(\vartheta_2)$, $X(\vartheta_3)$, $Z(\vartheta_4)$, $X(\vartheta_5)$, $Z(\vartheta_6)$ must be evaluated repeatedly ("on line"). Finally, mixed constant-variable compositions must be evaluated repeatedly, too. We shall focus our attention on the possibility of parallel "on line" calculations. The first phase of calculation consists of parallel evaluation of the sin and the cos functions of the arguments ϑ_i, $i = 1, 2, 3, 4, 5, 6$, i.e. of 12 parallel processes. The remaining calculations consist merely of parallel compositions, which can be done in four phases:

- phase 1: $\quad \sin(\vartheta_1), \qquad \sin(\vartheta_2), \qquad \sin(\vartheta_3), \qquad \sin(\vartheta_4), \qquad \sin(\vartheta_5), \qquad \sin(\vartheta_6),$
  $\quad\quad\quad\ \cos(\vartheta_1), \qquad \cos(\vartheta_2), \qquad \cos(\vartheta_3), \qquad \cos(\vartheta_4), \qquad \cos(\vartheta_5), \qquad \cos(\vartheta_6)$

- phase 2: $\quad U_1 = T_1 \cdot Z(\vartheta_1), \qquad U_2 = T_2 \cdot X(\vartheta_2), \qquad U_3 = T_3 \cdot X(\vartheta_3),$
  $\quad\quad\quad\ U_4 = T_4 \cdot Z(\vartheta_4), \qquad U_5 = T_5 \cdot X(\vartheta_5), \qquad U_6 = T_6 \cdot Z(\vartheta_6)$

- phase 3: $\quad\quad\quad\quad V_1 = U_1 \cdot U_2, \qquad V_2 = U_3 \cdot U_4, \qquad V_3 = U_5 \cdot U_6$

- phase 4: $\quad\quad\quad\quad\quad\ W_1 = V_1 \cdot V_2, \qquad W_2 = V_3 \cdot T_7$

- phase 5: $\quad\quad\quad\quad\quad\quad\ \Omega(\vartheta, \lambda) = W_1 \cdot W_2$

Let's denote by d_i the duration of the i-th phase. The total value of the elapsed time will be $\sum d_i$ in the case of parallel evaluation, and $12d_1 + 6d_2 + 3d_3 + 2d_4 + d_5$ in the case of sequential evaluation. As only a homogeneous transformation is performed in each of the last four phases, $d_2 = d_3 = d_4 = d_5$. Thus the total values of the elapsed time are $\sum d_i = d_1 + d_2 + d_3 + d_4 + d_5 = d_1 + 4d_2$ for the parallel evaluation and $12d_1 + 6d_2 + 3d_3 + 2d_4 + d_5 = 12(d_1 + d_2)$ for the sequential evaluation. Except for any additional overhead due to parallel calculations $d_1 + 4d_2 < 12(d_1 + d_2)$, since $d_1 > 0$, $d_2 > 0$.

3 Inverse Kinematics

Obtaining a solution for the joint coordinates is of the utmost importance in robot manipulator control. We normally know where we want to move the manipulator in terms of $\Omega(\vartheta, \lambda)$ in (1) and we need to obtain the joint variables in order to make the move. One of the main differences between the direct and the inverse kinematic transformation is the number of solutions. We have always only one solution of the direct kinematic transformation, whereas several solutions appear in the case of the inverse. Approximate numeric methods provide "the nearest" solution only. We shall focus our attention to the closed form solution methods providing all of the solutions and we shall show that these solutions may be evaluated simultaneously. We shall demonstrate the principle by an example of a so called "quadratic" structure with 8 solutions. A paradigm kinematic equation in a symbolic form is

$$Z(\vartheta_1)a(\lambda_1)X(\vartheta_2)c(\lambda_2)X(\vartheta_3)c(\lambda_3)Z(\vartheta_4)X(\vartheta_5)Z(\vartheta_6) = \begin{pmatrix} & \tau_1 \\ [r_{ij}], & \tau_2 \\ & \tau_3 \\ 0, & 1 \end{pmatrix}$$

The following 3 subtasks (each with 2 solutions) are to be solved in order to get a complete set of $2^3 = 8$ solutions:

$$\text{trans}\{Z(\vartheta_1)a(\lambda_1)y(\delta)\} = \text{trans}\{a(\tau_1)b(\tau_2)\} \tag{2}$$
$$\text{trans}\{X(\vartheta_2)c(\lambda_2)X(\vartheta_3)c(\lambda_3)\} = \text{trans}\{b(\delta)c(\tau_3)\} \tag{3}$$
$$Z(\vartheta_4)X(\vartheta_5)Z(\vartheta_6) = X(-\vartheta_3 - \vartheta_2) \cdot Z(-\vartheta_1) \cdot [r_{ij}] \tag{4}$$

The equation (2) provides a system of two scalar goniometric equations with unknowns ϑ_1, δ and given constants $\lambda_1, \tau_1, \tau_2$. The equation (3) provides a system of two scalar goniometric equations with unknowns ϑ_2, ϑ_3 and given constants $\delta, \lambda_2, \lambda_3, \tau_3$. The last equation is an orientation equation which provides a system of 9 mutually dependent scalar goniometric equations with unknowns ϑ_4, ϑ_5 and ϑ_6.

We can summarize the evaluation procedure as follows:

$$\delta^\circ = \pm\sqrt{\tau_1^2 + \tau_2^2 - \lambda_1^2}$$
$$\vartheta_1^\circ = \text{atan2}(\lambda_1\tau_2 - \delta^\circ\tau_1, \ \lambda_1\tau_1 + \delta^\circ\tau_2)$$

$$K = \lambda_2^2 - \lambda_3^2 + \delta^{\circ 2} + \tau_3^2 \qquad L = -\lambda_2^2 + \lambda_3^2 + \delta^{\circ 2} + \tau_3^2 \qquad M = \lambda_2^2 + \lambda_3^2 - \delta^{\circ 2} - \tau_3^2$$

$$N^\circ = \pm\sqrt{4\lambda_2^2\lambda_3^2 - M^2}$$
$$\vartheta_2^{\circ\circ} = \text{atan2}(-K\delta^\circ + N^\circ\tau_3, \ K\tau_3 + N^\circ\delta^\circ)$$
$$\vartheta_2^{\circ\circ} + \vartheta_3^{\circ\circ} = \text{atan2}(-L\delta^\circ - N^\circ\tau_3, \ L\tau_3 - N^\circ\delta^\circ)$$

$$[s_{ij}^{\circ\circ}] = X(-\vartheta_3^{\circ\circ} - \vartheta_2^{\circ\circ}) \cdot Z(-\vartheta_1^\circ) \cdot [r_{ij}]$$
$$s^* = \pm 1$$
$$\vartheta_4^{\circ\circ*} = \text{atan2}(s^* s_{13}^{\circ\circ}, \ -s^* s_{23}^{\circ\circ})$$
$$\vartheta_5^{\circ\circ*} = \text{atan2}(s^*\sqrt{1 - s_{33}^{\circ\circ 2}}, \ s_{33}^{\circ\circ})$$
$$\vartheta_6^{\circ\circ*} = \text{atan2}(s^* s_{31}^{\circ\circ}, \ s^* s_{32}^{\circ\circ})$$

Hence, as a consequence of mutual dependences of the three subtasks we obtain 2 solutions for ϑ_1, 4 solutions for ϑ_2, ϑ_3 and finally, 8 solutions for the last three joint variables $\vartheta_4, \vartheta_5, \vartheta_6$. The elements of all 8 solutions of inverse kinematics thus form a tree-like parallel scheme.

If we use a parallel processing machine with at least 8 processors, all the solutions of the inverse kinematics can be obtained simultaneously. This information may be effectively used to make some tactical decisions on higher levels of the robot control system. For example a "short jump" from one arm configuration to the other one (during the cartesian positioning) requires at least two solutions to be available at the moment.

The benefits of this kind of the parallel processing implementation are rather qualitative, having no influence on the overall duration of the computation. We shall now examine the potentials for decreasing the duration of the computation by means of a parallel processing.

It can easily be found that it is not necessary to complete one subtask in order to obtain the data necessary to start the next subtask. Thus, there is a potential for speeding up the computation of inverse kinematics. Furthermore, various calculations in the subtasks can be started in parallel, as their input data is available at the same instant of time. After a thorough examination, we can divide each of the three subtasks into several steps that must be done on a serial basis, while some of the calculations in each of the steps could possibly be done in parallel. The decomposition of the computational algorithm into the individual steps is not unique and can therefore be optimized to achieve the minimal duration of computation and perhaps also an evenly distributed usage of individual processors.

Here we present one possible decomposition:

Subtask I   **S1:** δ

   S2: $e_1 = \lambda_1 \tau_2 - \delta \tau_1, \quad e_2 = \lambda_1 \tau_1 + \delta \tau_2$

   S3: $e_3 = 1/(\tau_1^2 + \tau_2^2)$

   S4: $\sin(\vartheta_1) = e_1 e_3, \quad \cos(\vartheta_1) = e_2 e_3$

   S5: ϑ_1

Subtask II   **S1:** $e_4 = \delta^2 + \tau_3^2$

   S2: $K, \quad L, \quad M$

   S3: $e_5 = [1/(2\lambda_2)]/e_4, \quad N$

   S4: $-K\delta + N\tau_3, \quad K\tau_3 + N\delta, \quad e_6 = -L\delta - N\tau_3, \quad e_7 = L\tau_3 - N\delta$

   S5: $\sin(\vartheta_2 + \vartheta_3) = e_5 e_6, \quad \cos(\vartheta_2 + \vartheta_3) = e_5 e_7$

   S6: $\vartheta_2, \quad \vartheta_2 + \vartheta_3, \quad \vartheta_2 = (\vartheta_2 + \vartheta_3) - \vartheta_2$

Subtask III   **S1:** $\sin(\vartheta_1)\sin(\vartheta_2 + \vartheta_3), \ldots \quad matrix \ X(-\vartheta_3 - \vartheta_2) \cdot Z(-\vartheta_1)$

   S2: $matrix \ [s_{ij}]$

   S3: $\vartheta_4, \quad \vartheta_5, \quad \vartheta_6$

Closer examination of the solution steps reveals that the subtask II can be started just after the step **S1** of the subtask I. The subtask III can be started only after finishing both the step **S4** of the subtask I and the step **S5** of the subtask II. This mutual dependence of subtasks is exemplified in the Table 1. The whole computation can be divided into 3 serial phases **P1,P2,P3** (corresponding to rows of the table), whereas the computation in each of these phases is done in 1 to 3 parallel paths (corresponding to columns of the table). Going still deeper, the solution steps **S1**, ... in each of the paths are realized sequentially, but a great deal of the work in individual steps can be fulfilled in parallel. The nature of this algorithm reveals that it can be at best realized on a parallel machine with alternating sequential/parallel processing.

It is evident that the duration of each phase or step is equal to the duration of the most involved, "longest" path of the phase or step. Summarizing these maximal counts of mathematical operations through all members of a set of the solution steps, we get the

Table 1: Mutual Dependence of Subtasks I,II,III

Phase	Subtask I	Subtask II	Subtask III
P1	**S1** (0,1,0,1,2)		
P2	**S2-S4** (0,0,1,2,1)	**S1-S5** (0,1,0,4,4)	
P3	**S5** (1,0,0,0,0)	**S6** (1,0,0,0,0)	**S1-S3** (1,1,0,5,3)

numbers given in parentheses after the respective set of solution steps in the table field. The numbers correspond from left to right to the counts of the operations atan2 call, sqrt call, division, multiplication, and addition/subtraction.

Looking at the table data, one can easily see that the most involved of all phases are paths situated in the diagonal elements of the table. Assuming that each element of the matrix $[s_{ij}]$ is being evaluated in a dedicated processor and summing the contributions for these elements, we get the cost of a parallel evaluation as a 5-tuple (1,3,0,10,9). On the contrary, considering totally sequential evaluation, we obtain the 5-tuple (6,3,2,38,25).

The relative speedup thus obtained depends on the relative durations of mathematical operations under consideration, which are different for varied processors. To further exemplify the result we have considered as an example a hypothetical parallel multi-processor architecture with individual processors based on well-known Intel $80386 + 80387$ and Intel 80486 microprocessors. Neglecting the necessary overhead due to the parallel method of computation, we obtained in both cases the relative speedup of approximately 3.2.

4   Conclusion

It has been shown that both the direct and the inverse kinematic transformations may be considered as parallel processes. A parallel evaluation of the kinematic transformation can be implemented on several different levels. On the highest level it renders simultaneously all the solutions of the inverse kinematics, which qualitatively improves the robot control system. Two lower levels are characterized by the parallel evaluation of the (parts of) the transformation subtasks and the parallel evaluation of the quantities computed within the subtasks. Making use of these levels gives the speedup with a maximum of approximately 3. Considering even lower levels is beyond the scope of the robotic essence of the problem.

The authors are grateful to the Slovak Grant agency for science (grant No. 2/999467/93) for partial supporting of this work.

References

Dobrovodský, K. (1992). Automatic Generation of Robot Inverse Kinematics. In: Kybernetika a informatika, Bratislava, 2/3

Kurdel, P. (1992). Analysis of the Solution Methods of the Prototype XbXb=bc - a Typical Subtask of the Inverse Kinematics Problem. In: Kybernetika a informatika, Bratislava, 2/3

Rieseler, H., Schrake, H., Wahl, F.M. (1991). Fourth and Second Order Solutions for Robots with Planar Joints Sets. In: 14th IASTED International Symposium Manufacturing and Robotics, June 25-27, Lugano, Switzerland

Robot Arm Modelling and Control

S. Uran, K.Jezernik
Laboratory of Industrial Robotics
Faculty of Technical Sciences - ERI, University of Maribor,Slovenia

Abstract: At the Faculty of Technical Sciences Maribor the course in robot modelling and control is held within the undergraduate and graduate course for students of Electrotechnics. Lectures and exercises of robot modelling and control are supplemented with computer simulations and laboratory experiments. In the paper the equipment of the Laboratory for Robotics, the concept of undergraduate student exercises and the research projects for graduate students are described. Along with the description of exercises and projects some laboratory results are represented.

1. Robotic laboratory mechanism

At the Faculty of Technical Sciences Maribor the course in robot modelling and control within the undergraduate and graduate course is held for students of Electrotechnics. Lectures and exercises are supplemented with computer simulations and laboratory experiments. The object for modelling and control is a robotic laboratory mechanism. The robotic laboratory mechanism has SCARA configuration (3 D.o.F.), but for educational purposes it is mainly used with 2 D.o.F., as shown on Fig.1.

Figure1: Robotic laboratory mechanism

From the control point of view such configuration (2 D.o.F.) shows nonlinear coupling between the joints through the full inertia matrix and Coriolis and centripetal forces, while no gravity force exists.

Each joint of the robotic laboratory mechanism is driven with DC or AC electrical motor with gear-box. The type of motor to drive robotic laboratory mechanism could be chosen, but usually we use DC motor to drive the first joint and AC motor to drive the second joint. The gear ratio used in each joint is 100 and is quite high.

The controller of the robotic laboratory mechanism is based on a personal computer (PC) and its transputer card [Terbuc et.al.1993] and is oriented toward experiments in robot control. This means that it supports many different robot control algorithms and an easy implementation of new ones.

2. Undergraduate Course

Undergraduate students build a dynamic model of the 2 D.o.F. robotic laboratory mechanism in the symbolic form. For the derivation of a dynamic model they use Lagrange's approach based on the energy of the mechanism. After a successful derivation of the dynamic model of the mechanism they join obtained model with the model of the gears and with a model of electric motors.

In order to verify the dynamic model of the laboratory mechanism students perform simulations. For this purpose they use block oriented computer simulation program PADSIM developed on the Faculty of Technical Sciences Maribor.

Then are undergraduate students directed towards the design of standard robot control algorithms, such as: Cascade Control and Computed Torque Control in the joint space. The students are advised to take into account existing equipment of the robotic laboratory mechanism controller. In this way they become familiar with the components of the controller and make their simulation results as close to the laboratory experiments as possible. They implement a motor current control loop in the case of all robot control algorithms. In the case of Cascade Control the students implement after the motor current control loop also speed control loop with PI controller and position control loop with P controller. The design of speed and position control loops they perform by the well known procedures in the frequency domain (with the usage of Bode diagram or root locus), like the method of symmetric optimum, pole placement etc., or in the time domain (with the usage of time response or Integral of square error criterion). In the case of Computed Torque Control they implement after the motor current control loop inverse robot dynamic model for external linearization and PD position controllers or state controllers.

The students extend their simulation model with task generators (circles,lines) in the task space and (PTP) in joint space and transformations (robot kinematic models) from task space into joint space and vice versa. Obtained simulation result of such model for a circular task is a circle in the x-y plane. Through simulation results acheived control behavior could be verified.

After a successful verification of the robot model and the design of robot control algorithm the students implement their controllers in the transputer based controller of the laboratory mechanism. Because the laboratory experiments are always connected with a risk of damage the undergraduate students are not completely free at the implementation of controllers. In order to reduce possible damage of the laboratory mechanism standard robot control algorithms for instance Cascade controller or Computed Torque Controller are already implemented in the controller of the laboratory mechanism and the adjustment of their parameters is possible. Through the adjustment of parameters of robot control algorithms students could influence the controller design.

Position, speed and motor current could be measured by the controller and shown on the screen. Monitoring of the laboratory mechanism's tip movement is also performed on the basys of the plots made by a pencil mounted on the tip of the laboratory mechanism. In contrast to very expensive equipment for measurement of the robot tip movements in Cartesian space (3D) a decision for 2 D.o.F. mechanism and plotting of tip movement is a proper and cheap solution well suited for educational needs. In addition to this Benchmark test for control algorithms consist mostly from circles and lines in the task plane. An example of circles plotted by the laboratory mechanism during adjustment of the Cascade Controller parameters is shown in Fig.2.

The figure shows great influence of the position controller parameter on the performance of the cascade control. The lower is the value of position controller the greater is deviation of plotted circle from the reference one.

3. Graduate Courses

Graduate students of robotics work on different problems during a research project. They try to obtaine detailed robot models, to suggest new robot control algorithms, to evaluate performances of advanced robot control algorithms, to solve implementation problems of the advanced robot control algorithms etc. According to their needs they can modify existing programs of the laboratory mechanism controller and add their own programs written in the C language. No industrial robot controller exists that would support such program changes, therefore the development of the laboratory mechanism controller was a necessity.

On the field of robot control an implementation and an evaluation of performances of different robot control algorithms based on the rigid body robot model was performed recently. In the controller of the robotic laboratory mechanism centralized and decentralized robot control algorithms were implemented.

The most representative of the centralized robot control algorithms is the Computed Torque Method. With this method nonlinear robot dynamic model could be globally linearized and decoupled, and the dynamic behavior of each robot joint could be adjusted by a linear controller. Due to great computation effort needed for an implementation of the Computed Torque Method it is still not widely used for control of industrial robots today.

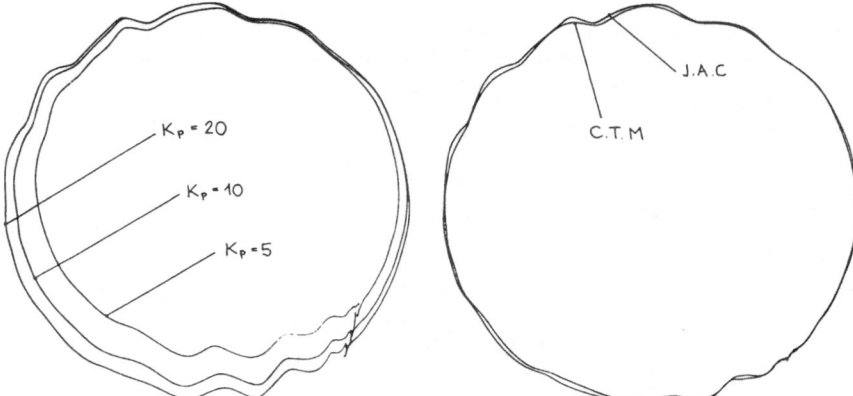

Figure2: Circles plotted with different Cascade Controller parameters

Figure3: Circle plotted Joint Acceleration Controller and Computed Torque Method

Decentralized control algorithms are widely used for control of industrial robots today, because for their implementation low computational effort is needed. The most widely used decentralized robot control algorithm is the Cascade Control. Recently new advanced decentralized robot control algorithms [Nakao et.al.1987, Komada and Ohnishi 1989] were proposed. These control algorithms are more robust against plant parameter variations and unmodeled dynamics. They are observer based and could by the use of observer effectivelly linearize and decouple nonlinear robot dynamic model. Therefore their performances are comparable to the performance of the Computed Torque Method.

In the controller of the robotic laboratory mechanism along with the Cascade Control and Computed Torque Control also some advanced decentralized robot and centralized control algorithms were implemented.

Among the advanced decentralized control algorithms Joint Acceleration Controller was implemented. Joint Acceleration Controller is based on a disturbance observer. According to the consideration that plant parameter variations, nonlinear joint coupling, Coulomb friction and unmodeled dynamics could be considered as disturbances of one joint, they could be observed by the disturbance observer. Therefore observed disturbances could also be compensated. By such compensation the resulting joint model is decoupled and linear with constant parameters.

In order to acheive the most suitable implementation of the disturbance compensation the motor current control loop should be implemented. The motor current control loop enables to control motor current independent of motor speed and to consider the reference current proportional to the actual current. Because the motor current loop is very fast in comparison to the rest of the control plant it could be considered as fast enough for disturbance compensation.

If the compensation of the disturbances is succesfully implemented actual acceleration is expected to be equal to the reference acceleration of the joint. Of course, actual acceleration could never become equal to the reference acceleration. Reasons for this are disturbance observer's own dynamics and already considered dynamics of the motor current control loop.

Through such decision all the differences due to different motor types (DC,AC) of the robotic laboratory mechanism are solved in the analog current control loop. So the rest of the Joint Acceleration Controller which is implemented digitaly is the same for both joints. In order to implement disturbance observer digitaly it was discretized and designed in its discrete form. The discrete form of the disturbance observer was derived from the simplified, decoupled and linear model with constant parameters by the method of step invariance. The disturbance was modeled as a constant. Disturbance observer is represented with equation 1:

$$\mu(K+1) = (1 + h.\tfrac{T}{J}).\mu(K) - h.\tfrac{K_M.T}{K_{MI}.J}.i_{ar}(K) + h^2.\tfrac{T}{J}.\dot{q}(K)$$
$$\widehat{M}_d(K) = \mu(K) + h.\dot{q}(K) \tag{1}$$

In the equation 1 $\dot{q}(K)$ denotes joint speed, $i_{ar}(K)$ denotes reference motor current, $\widehat{M}_d(K)$ denotes estimated disturbance torque of the joint. h represents observer parameter.

The control law is represented with equation 2:

$$i_{ar}(K) = \frac{J}{K_M}.\{\ddot{q}_r(K) + K_p.[q_r(K) - q(K)] + K_v.[\dot{q}_r(K) - \dot{q}(K)]\} + \frac{1}{K_M}.\widehat{M}_d(K) \tag{2}$$

In order to compute the Joint Acceleration Controller 8 multiplications and 6 additions (substractions) are needed per joint. Constant parameters are calculated off-line in order to reduce computation time for the control algorithm. After the implementation of the Joint Acceleration Controller with a sampling time of 1 ms a greate reserve of computation time remained. Experimental results in a case of the circular movement with implementaticn of the Joint Acceleration Controller in comparison with experimental results obtained with implementation of Computed Torque Method are shown in Fig.3.

Next the Acceleration Controller in Cartesian external coordinates (task space) was implemented. The control law is given with equation 3:

$$i_{ar}(K) = \Phi\{\ddot{X}_r(K) + K_p.[X_r(K) - X(K)] + K_v.[\dot{X}_r(K) - \dot{X}(K)]\} + \frac{1}{K_M}.\widehat{M}_a(K) \tag{3}$$

Φ represents transformation of accelerations from external Cartesian coordinates into joint coordinates and further into reference current, X represents x or y external coordinates.

The computation effort needed for the calculation of the acceleration controller is greater than for Joint Acceleration Controller, because in addition to the before mentioned computations computations of direct and inverse kinematic models (with transcendental functions) are also needed. A sampling time of 1 ms with our transputer based laboratory mechanism controller was still acheived. Detailed description of this algorithm and its implementation is in [Terbuc et.al. 1993]. Experimental results in a case of the circular movement with implementation of the Acceleration Controller in comparison with experimental results obtained with implementation of Computed Torque Method are shown in Fig.4.

Acceleration controller in Cartesian coordinates offers an advantage of easier implementation of control laws based on external sensors, such as force sensors and cameras. Temporary an implementation of the hibrid position force controller based on the acceleration controller is performed. The control law for hibrid position force controller is represented with equation 4:

$$\tilde{i}_{ar}(K) = \Phi\Big[\{\ddot{X}_r(K) + K_p.[X_r(K) - X(K)]\}.S + K_v.[\dot{X}_r(K).S - \dot{X}(K)]\Big] + \tfrac{1}{K_M}.\widehat{M}_d(K)$$
$$i_{ar}(K) = \tilde{i}_{ar}(K) + \Phi\Big[\tfrac{(1-S)}{M}.[F_r(K) - F(K)]\Big] \tag{4}$$

It demands only 3 multiplications and 2 additions more than Acceleration Controller therefore it is expected that a sampling time of 1 ms would be acheived.

In our Laboratory for Robotics adaptive control algorithms like model reference or self-tuning control algorithms were not explored, because the majority of them is also based on linearized models. Because of the need for adaptation of the control algorithms a research of neural networks was started. Neural network controller which was implemented is a modification of the Computed Torque Method. In this controller two neural multilayer networks with one hidden layer and back propagation learning procedure are used.

58

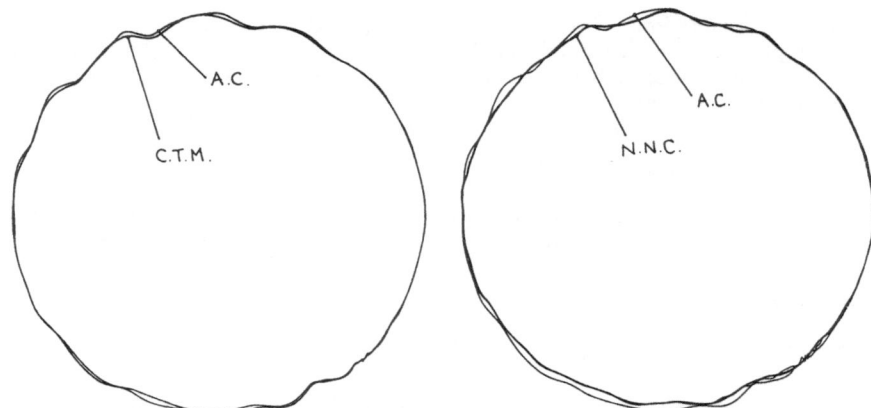

Figure4: Circle plotted with Acceleration Figure5: Circle plotted with Neural Net-
Controller and Computed Torque Method work and Acceleration Controller

The first neural network is used for the calculation of robot inertia matrix and the second
is used for the compensation of Coriolis, centripetal or gravitational torques. Detailed
description of this algorithm and its implementation is in [Safaric et.al.1993]. Experimental results in a case of the circular movement with implementation of the Neural network
controller in comparison with experimental results obtained with implementation of Acceleration Controller are shown in Fig.5.

All control approaches were implemented on a transputer based laboratory mechanism
controller. Transputer network of this controller consists of four transputers. Among
them the transputer called T3 executes actual control algorithm. When needed, computation parallelization of transputers T3 and T4 could be acheived.

A comparison of experimental results of different control approaches shows that with
advanced decentralized control approaches good experimental results could be obtained.
Experimental results obtained with neural network controller show the greatest similarity
to the circular shape of the reference trajectory.

References

An C.H., Atkeson C.G., Griffiths J.D.,Hollerbach J.M.(1989): Experimental Evaluation
of Feedforward and Computed Torque Control. IEEE Trans. on Robotics and Automa-
tion, Vol.5, No.3,pp."368-372"
Craig J.J.(1986): Introduction to Robotics, Mechanics and Control. Addison Wesley
Publishing Company,Reading, Massachusetts
Khosla P.K., Kanade T.(1989): Real-time implementation and evaluation of Computed
Torque Scheme. IEEE Trans. on Robotics and Automation, Vol.5, No.2,pp."245-253"
Komada S., Ohnishi K.(1989): Control of Robotic Manipulators by Joint Acceleration
Controller. Proc. of Int. Conf. on Ind.Electronics, Control and Instrumentation 1989,
Philadelphia,pp."623-628"
Nakao M., Ohnishi K., Miyachi K.(1987): A robust decentralized joint control based
on interference estimation . Proc. of IEEE Int. Conf. on Robotics and Automation,
Raleigh,pp."326-331"
Stute G.(1981): Regelung an Werkzeug Maschinen. Carl Hanser Verlag München
Šafarič R., Jezernik K., Terbuc M.(1993): Transputer Based Neural Network Controller
for a two D.o.F. SCARA Mechanism Proc. of Int. Symp. on Mathem. and Intel. Models
and Syst. simulation '93 , Brusselex, Vol.2, pp."15-18"
Terbuc M.,Zimšek A., Jezernik K.(1993): Transputer based multiprocessor robot controller. Proc. of Int. Conf. on DMMI'93, Bled Slovenia, pp."445-451"
Uran S., Jezernik K., Prosenak D.(1993): An Implementation of the Robot Joint Acceleration Controller. Proc. of Int. Conf. on DMMI'93, Bled Slovenia, pp."437-444"

Dynamic Model of a Mobile Robot for Analyzing End-effector Deviations Caused from Ground Unevenness

G. Elsbacher

Institute for Handling Devices and Robotics
University of Technology, Vienna, Austria

Abstract: This paper presents a new model for a mobile robot. With the aid of past models it is impossible to analyze the dynamic response of the end effector by appearance of some disturbances such as ground unevenness. The model presented in this paper includes six degrees of freedom (6 DOF) for the robot and 6 DOF for the vehicle. A robot of type "manutec r3" with six degrees of freedom and a vehicle with two driven wheels and one supporting wheel (produced from Fa. Schindler) is used as a specimen for the investigations. A different number of DOF's in the velocity level (5 DOF) and position level (6 DOF) exists because of nonholonomic connections between tires and ground. The mechanical substituting system of the robot is a multi-body-system with rigid bodies having its own 6 DOF.

1. Introduction

By means of mobile robots many new applications for robots are being used - like welding of large works. This task is principly solvable with a portal robot, but in reality it isn't practicable because of the limited radius of operation. Combination of motion and location brings many new control tasks, like optimal motion planning or reducing end effector deviations caused from ground unevenness. For consideration of disturbances caused by ground unevenness in controller design it is necessary that the mathematical model of the system has additional degrees of freedom for the space-motion of the vehicle. In this paper the generation of the dynamic equations with the following properties is presented:

- Vehicle is modelled with six degrees of freedom (Cartesian position and orientation).
- Elasticities and damping of the vehicle are concentrated by the wheel- ground connection in form of elastic and absorbing elements.
- Dynamic equations are generated with a recursive Euler-method.

2. Generate the dynamic equations

2.1 Mechanical models
2.1.1 Mechanical model of the robot

The mechanical model of the robot is a multibody-system with rigid bodies, which is described in the bibliography (Türk, 1988), having the following properties:

- The robot consists of twelve rigid bodies (six rotors, and six links)
- Joints are numbered from one to six; Pedestal is a part of the vehicle
- There exists only one (rotational) degree of freedom between two following bodies
- Generalized coordinates are the relative angles between the joints.

2.1.2 Mechanical model of the vehicle

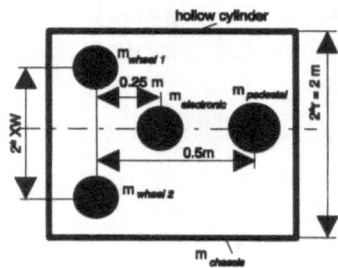

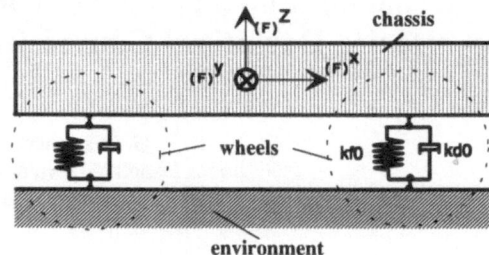

Figure 1: model of the vehicle   Figure 2: model of the tyres

Because of the difficult geometry of the vehicle following simplifications are defined (Rämisch, 1991):

- Wheels, electronics and pedestal are considered as point masses (Fig. 1)
- Chassis is considered as a thin-walled hollow cylinder with radius $r=1m$ (Fig. 1)
- Tyres of the wheels are modelled as elastic and absorbing elements with damping coefficient K_d and spring coefficient K_f (Fig. 2).

2.2 Dynamic equation of the overall system

The dynamic equations of motion of multi-body-system can be written as follows:

$$\mathbf{M(q)} * \ddot{\delta} + \mathbf{k}(\dot{\delta}, \mathbf{q}) + \mathbf{g(q)} = \tau \qquad (1)$$

where:

$\mathbf{M(q)}$ -$\langle 11 \times 11 \rangle$ symmetric and positive definite inertia matrix

$\mathbf{q}$ -$\langle 12 \times 1 \rangle$ vector of the angles of joint rotations and vehicle positions and orientations

$\dot{\delta}, \ddot{\delta}$ -$\langle 11 \times 1 \rangle$ vectors of the velocity state variables (minimal coordinates) and accelerations (output variables of the system)

$\mathbf{k}(\dot{\delta}, \mathbf{q})$ -$\langle 11 \times 1 \rangle$ vector of centrifugal and coriolis moments or forces

$\mathbf{g(q)}$ -$\langle 11 \times 1 \rangle$ vector of gravity moments or forces

τ -$\langle 11 \times 1 \rangle$ vector of impressed torques or forces.

A practical method to generate and calculate equation (1) for systems with this size (seven bodies) is the EULER-method (Bremer, 1988):

$$\sum_{k=1}^{7} (\mathbf{M}_k * \ddot{\delta} + \mathbf{k}_k + \mathbf{d}_k) = 0 \qquad (2)$$

with

$$\mathbf{M}_k = \mathbf{J}_{k,T}^T * m * \mathbf{J}_{k,T} + \mathbf{J}_{k,R}^T * \mathbf{I}_k * \mathbf{J}_{k,R}$$

$$\mathbf{k}_k = \mathbf{J}_{k,T}^T * m *_k \ddot{\mathbf{p}}_v^k + \mathbf{J}_{k,R}^T * \mathbf{I}_k *_k \dot{\omega}_v^k + \mathbf{J}_{k,R}^T * \tilde{\omega} * \mathbf{I}_k * \omega$$

$$\mathbf{d}_k = -\mathbf{J}_{k,T}^T *_k \mathbf{f}_{e_k} - \mathbf{J}_{k,R}^T *_k \mathbf{l}_{e_k} = \mathbf{g}_k - \tau_k$$

For calculation of the equation (2) it is necessary for all bodies to compute the following terms:

angular velocities $(_k\omega^k)$ from the centres of mass, terms of accelerations depending on velocities $(_k\ddot{\rho}_v^k$ and $_k\dot{\omega}_v^k)$, one Jacobean-matrix for linear motion $(_kJ_T)$ and one for rotation $(_kJ_R)$, the impressed forces (f_{e_k}) and torques (l_{e_k}) and the constant tensor of inertia (I_k).

The Jacobean-matrixes $(_kJ_T$ and $_kJ_R$) project the degrees of freedom of the bodies to the minimal coordinates (velocity level).

$$_kJ_{k,T} = \left[\frac{\delta_k\dot{\rho}^k}{\delta_{\hat{F}}v_y} \quad \frac{\delta_k\dot{\rho}^k}{\delta_{\hat{F}}v_z} \quad \frac{\delta_k\dot{\rho}^k}{\delta_I\omega_x} \quad \frac{\delta_k\dot{\rho}^k}{\delta_I\omega_y} \quad \frac{\delta_k\dot{\rho}^k}{\delta_I\omega_z} \quad \frac{\delta_k\dot{\rho}^k}{\delta\dot{\Theta}_1} \quad \cdots \quad \frac{\delta_k\dot{\rho}^k}{\delta\dot{\Theta}_6} \right]$$

$$_kJ_{k,R} = \left[\frac{\delta_k\omega^k}{\delta_{\hat{F}}v_y} \quad \frac{\delta_k\omega^k}{\delta_{\hat{F}}v_z} \quad \frac{\delta_k\omega^k}{\delta_I\omega_x} \quad \frac{\delta_k\omega^k}{\delta_I\omega_y} \quad \frac{\delta_k\omega^k}{\delta_I\omega_z} \quad \frac{\delta_k\omega^k}{\delta\dot{\Theta}_1} \quad \cdots \quad \frac{\delta_k\omega^k}{\delta\dot{\Theta}_6} \right]$$

(3)

where:

$_k\omega^k$ -rotational velocity of body k referred to the coordinate frame k

$_k\dot{\rho}^k$ -linear velocity of body k referred to the coordinate frame k

2.3 Kinematics of the general system

The angular velocity can be calculated using the following recurrent equation:

$$_k\omega^k = A^{k,k-1} *(_{k-1}\omega^{k-1} + _{k-1}\Omega^k)$$ (4)

where:

$A^{k,k-1}$ -transformation matrix between body k and $k-1$

$_k\omega^k$ -absolute angular velocity of the body k referred to the coordinate frame k

$_{k-1}\Omega^k$ - relative angular velocity between body k and $k-1$ referred to coordinate frame $k-1$.

After deriving equation (4) we obtain the recurrent equation for the angular acceleration $_k\dot{\omega}^k$:

$$_k\dot{\omega}^k = A^{k,k-1} *(_{k-1}\dot{\omega}^{k-1} + _{k-1}\tilde{\omega}_{I,k-1} * _{k-1}\dot{\Omega}^{k-1})$$ (5)

where $_{k-1}\tilde{\omega}_{I,k-1}$ is the skew-symmetric tensor of the angular velocity (Bremer, 1989).

Perpendicular distances of the fulcrum from the line of action of the forces for the calculation of the linear accelerations are shown for the robot in Fig. 3 and for the vehicle in Fig. 4. With these definitions we obtain the following recurrent equation for the linear accelerations depending on velocities

$$_k\ddot{\rho}_v^k = A^{k,k-1} *(_{k-1}\ddot{\rho}_v^{k-1} + _{k-1}(\tilde{\omega}*\tilde{\omega})_{I,k-1} * _{k-1}t^k + 2*(_{k-1}\tilde{\omega}_{I,k-1})^2 * A^{k-1,k} * _k su^k +$$
$$_{k-1}(\tilde{\omega}*\tilde{\omega})_{k,k-1} * A^{k-1,k} * _k su^k)$$ (6)

with

$$_{k-1}t^k = _{k-1}so^{k-1} + A^{k-1,k} * _k su^k$$

62

$$_k\ddot{\rho}_v^k = \mathbf{A}^{k,k-1} *(_{k-1}\ddot{\rho}_v^{k-1} +_{k-1}(\tilde{\omega}*\tilde{\omega})_{l,k-1} *_{k-1}\mathbf{t}^k + 2*(_{k-1}\tilde{\omega}_{l,k-1})^2 * \mathbf{A}^{k-1,k}*_k\mathbf{su}^k +$$

$$_{k-1}(\tilde{\omega}*\tilde{\omega})_{k,k-1} * \mathbf{A}^{k-1,k}*_k\mathbf{su}^k)$$ (6)

with

$$_{k-1}\mathbf{t}^k =_{k-1}\mathbf{so}^{k-1} + \mathbf{A}^{k-1,k}*_k\mathbf{su}^k$$

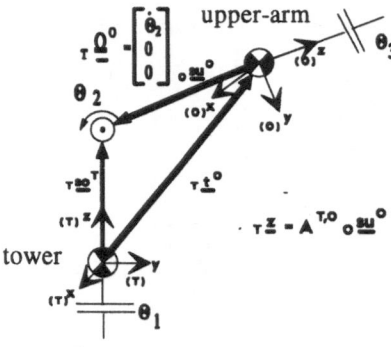

Figure 3: Vectors relating the link
"tower" with the link "upper-arm"

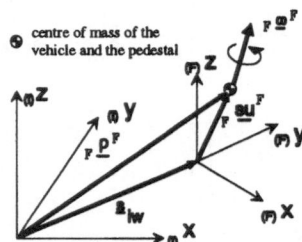

Figure 4: model of the tyres

2.4 Impressed torques and forces
2.4.1 Impressed torques and forces of the vehicle

Generally here are included forces of gravitation and forces (moments) transmitted by tyres. The second group of forces can be subdivided into driving forces, support forces, and forces (moments) coming from ground unevenness. Gravitation forces transformed into coordinate frame of the vehicle are given by:

$$_F\mathbf{f}_G^F = \begin{pmatrix} _F F_{Gx}^F & _F F_{Gy}^F & _F F_{Gz}^F \end{pmatrix}^T = \mathbf{A}^{F,I}*\begin{pmatrix} 0 & 0 & -m_g*g \end{pmatrix}^T$$ (7)

Since the centre of mass of the vehicle isn't in the centre of rotation, there are acting moments round the centre of rotation. Driving forces reduced on the centre of rotation are given :

$$\begin{pmatrix} _F F_y \\ _I M_z \end{pmatrix} = \begin{bmatrix} 1 & 1 \\ x_w & -x_w \end{bmatrix} * \begin{pmatrix} _F F_{v1} \\ _F F_{v2} \end{pmatrix}$$ (8)

where $_F F_{v1}$ and $_F F_{v2}$ are the driving forces of the wheels 1 and 2 referred to the coordinate frame $\tilde{F}$ (Elsbacher, 1992)

Additional to the driving forces there are "variable" support forces. One effect is to hold in Equilibrium the general system. Spoiling the vehicle (through ground unevenness) is the second effect of these forces. To calculate the effective forces we define the following simplification: Only forces that are active in the z-direction of the coordinate frame are considered. That means, effect of forces in longitudinal direction of the vehicle during driving up to the obstacle are ignored. Obstacles and ground unevenness have the same effects to the model as a "lift", which is working under each tyre, and pressing it up (Fig. 6).

To generate the expressions for the forces it is necessary to make some reflections about the kinematics. The distance between wheel 1 and the centre of the world coordinate frame is given as (Fig. 5):

$$_I\mathbf{x}_{R1} = {}_I\mathbf{x}_F + \mathbf{A}^{I,F} *_F \mathbf{x}_{F,R1} \tag{9}$$

The spring shortening $_I\Delta l_1$ is given :

$$_I\Delta l_i = {}_Iz_{steuer\,i} - {}_Iz_{R\,i} \qquad , i = 1...3 \tag{10}$$

where $_Iz_{steuer\,i}$ is the "desired spoil" and $_Iz_{Ri}$ is the z-coordinate of the vektor $_I\mathbf{x}_{Ri}$. The relation on velocity level is given:

$$_I\dot{\Delta l}_i = {}_I\dot{z}_{steuer\,i} - {}_I\dot{z}_{R\,i} \qquad , i = 1...3 \tag{11}$$

The "spring-absorber" - law provides the forces (in z-direction):

$$_I F_{z_i} = k_{f_i} *_I \Delta l_i + k_{d_i} *_I \dot{\Delta l}_i \quad , i = 1..3 \tag{12}$$

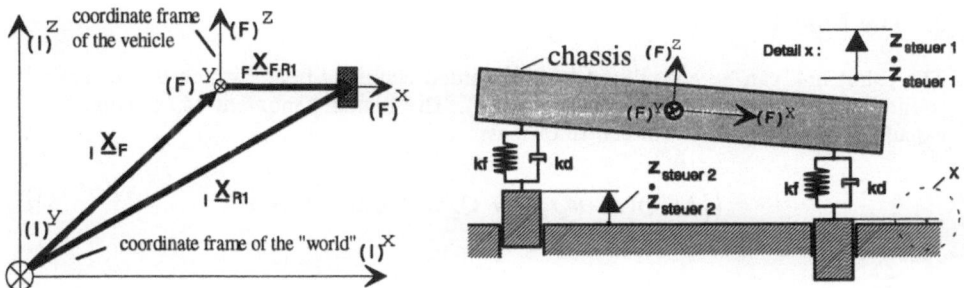

Figure 5: Geometrical relations at wheel 1   Figure 6: Modelling the spoiling of the vehicle

3. Conclusion

This paper presents a new mathematical model for a mobile robot. This model makes it possible to realise the spoiling of the vehicle. A robot of type "manutec r3" and a vehicle with two driven wheels and one supporting wheel is used as a specimen for the investigations. The tyres of wheels are modelled as elastic and absorbing elements. The mechanical substituting system of the robot is a multi-body-system with rigid links. The calculations of the dynamic equations are shown. To analyze the dynamic response of the end effector by this generated model any disturbance under the wheels can be impressed.

4. References

Bremer, H. (1988). *Dynamik und Regelung mechanischer Systeme*. B.G. Teubner , Stuttgart.

Considerations for the Construction of Lightweight Robots

K. Desoyer[1], I. Caushi[2] and P. Kopacek[2]
[1]Institute for Mechanics
[2]Institute for Handling Devices and Robotics
University of Technology, Vienna, Austria

Abstract. To reduce the mass of a robot link having the same deflection by the same length under the same loading, there exist the possibilities of using lighter materials and/or other geometric form, both by obtaining the needed rigidity. Possibilities of minimizing the weight of robot links, which are bending under static and dynamic loadings, by constructive arrangements and suitable form giving are here investigated.

1. Introduction

Arrangements to reduce the robot-to-payload mass ratio are being intensively investigated (Desoyer and Kopacek, 1992), (Çaushi, 1993). Two essential ways are to distinguish: 1. The use of lightweight constructions, which leads to elastic robots and 2. the use of new construction forms (Walser, 1991) and materials (Urbanek, 1988). Here are examined possibilities to these aim by giving geometric suitable form.

2. The Model

The robot link can be considered as a one-sided clammed beam, which is accelerated round an inertial fixed point (Walser, 1991). The bending moment $M_y(x)$ (Fig. 1) is calculated by the theorem of centre of mass

$$[\int_x^l \rho A(\xi)\xi d\xi + m_2 r_2]\ddot{\psi} = Q_z(x) - \int_x^l \rho g A(\xi) d\xi - m_2 g \tag{1}$$

and the theorem of moments

$$I_y\ddot{\psi} = -M_y(x) + xQ_z(x) - \int_x^l \rho g A(\xi)\xi d\xi - r_2 m_2 g \tag{2}$$

with the moment of inertia

$$I_y = \int_x^l \rho A(\xi)\xi^2 d\xi + \int_x^l \rho J_y(\xi) d\xi + I_{2,C_2,y} + m_2 r_2^2 \tag{3}$$

With equation (3) follows from (1) and (2) the bending moment $M_y(x)$:

$$M_y(x) = -[\int_x^l \rho A(\xi)(\xi-x)\xi d\xi + \int_x^l \rho J_y(\xi) d\xi + I_{2,C_2,y} + m_2 r_2(r_2-x)]\ddot{\psi}$$
$$- \int_x^l \rho g A(\xi)(\xi-x) d\xi - (r_2-x)m_2 g \tag{4}$$

For the rotation round the vertical axis there is an analog result:

$$M_z(x) = -[\int_x^l \rho A(\xi)(\xi-x)\xi d\xi + \int_x^l \rho J_z(\xi)d\xi + I_{2,C_2,z} + m_2 r_2(r_2-x)]\ddot{\theta} \qquad (5)$$

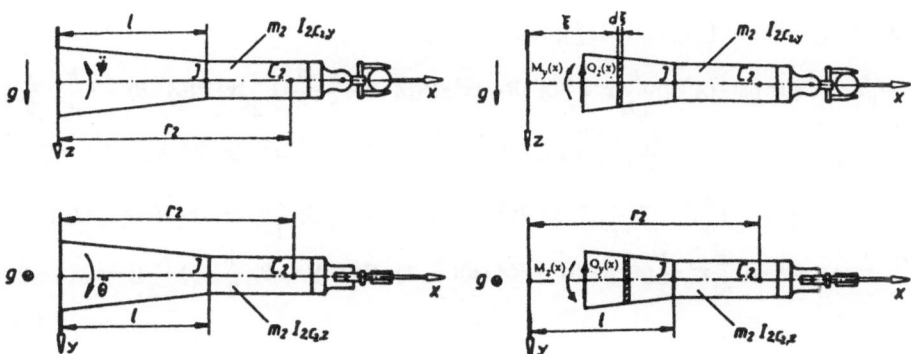

Fig. 1 Mechanical model of the robot arm

For the **elliptic hollow cross-section** we consider linear decreasing of the dimensions:
$$a = A(1-k\,\xi/l) \quad \text{and} \quad b = B(1-k\,\xi/l) \qquad (6)$$

A and B denote respectively the great and small half-axis at the fixation position $\xi=0$.

Denoting the wall thickness with s, for the areal moments of inertia $J_y^E(\xi)$ and $J_z^E(\xi)$, the cross-sectional area $A^E(\xi)$ and the bending moments M_y^E and M_z^E yields:

$$J_y^E(\xi) = \frac{\pi}{4}s[A^2(A+3B)(1-\frac{k}{l}\xi)^3 - 3sA(A+B)(1-\frac{k}{l}\xi)^2 + s^2(3A+B)(1-\frac{k}{l}\xi)-s^3] \qquad (7)$$

$$J_z^E(\xi) = \frac{\pi}{4}s[B^2(B+3A)(1-\frac{k}{l}\xi)^3 - 3sB(B+A)(1-\frac{k}{l}\xi)^2 + s^2(3B+A)(1-\frac{k}{l}\xi)-s^3] \qquad (8)$$

$$A^E(\xi) = \pi[ab-(a-s)(b-s)] = \pi s(a+b-s) = \pi s[(A+B)(1-\frac{k}{l}\xi)-s] \qquad (9)$$

$$M_y^E = -\ddot{\psi}[\rho\pi s\int_x^l [(A+B)(1-k\frac{\xi}{l})-s](\xi-x)\xi d\xi + I_{2,C_2,y} + m_2 r_2(r_2-x)$$
$$+\rho\frac{\pi}{4}\int_x^l [sA^2(A+3B)(1-\frac{k}{l}\xi)^3 - 3s^2A(A+B)(1-\frac{k}{l}\xi)^2 + s^3(3A+B)(1-\frac{k}{l}\xi)-s^4]d\xi] \qquad (10)$$
$$-\pi g\rho s\int_x^l [(A+B)(1-\frac{k}{l}\xi)-s](\xi-x)d\xi - m_2 g(r_2-x)$$

$$M_z^E = -\ddot{\theta}[\rho\pi s\int_x^l [(B+A)(1-k\frac{\xi}{l})-s](\xi-x)\xi d\xi +I_{2,C_2z}+m_2r_2(r_2-x)$$

$$+\rho\frac{\pi}{4}\int_x^l [sB^2(B+3A)(1-\frac{k}{l}\xi)^3 -3s^2B(B+A)(1-\frac{k}{l}\xi)^2 +s^3(3B+A)(1-\frac{k}{l}\xi)-s^4]d\xi]$$

(11)

The integration of (14) and (15) gives:

$$M_y^E(x) = -\ddot{\psi}[\pi\rho s[-(A+B)\frac{k}{12l}x^4 +\frac{1}{6}(A+B-s)x^3 +l^2[(A+B)(\frac{k}{3}-\frac{1}{2})+\frac{s}{2}]x+l^3[(A+B)(\frac{1}{3}-\frac{k}{4})-\frac{s}{3}]]$$

$$+I_{2,C_2y}+m_2r_2(r_2-x)+\frac{\pi\rho}{4}[\frac{sA^2}{4}(A+3B)[(\frac{k}{l})^3x^4+4(\frac{k}{l})^2x^3+2(\frac{k}{l})x^2-4x-lk^3-4lk^2-2lk+4l]$$

$$-s^2A(A+B)[-(\frac{k}{l})^2x^3+3(\frac{k}{l})x^2-3x+lk^2-3lk+3l]+\frac{s^3}{2}(3A+B)(\frac{k}{l}x-2x-lk+2l)+s^4(x-l)]]$$

$$-\pi g\rho s[-(A+B)\frac{k}{6l}x^3+\frac{1}{2}(A+B-s)x^2+l[(A+B)(\frac{k}{2}-1)+s]x+l^2[(A+B)(\frac{1}{2}-\frac{k}{3})-\frac{s}{2}]]-m_2g(r_2-x)$$

(12)

$$M_z^E(x) = -\ddot{\theta}[\pi\rho s[-(B+A)\frac{k}{12l}x^4 +\frac{1}{6}(B+A-s)x^3 +l^2[(B+A)(\frac{k}{3}-\frac{1}{2})+\frac{s}{2}]x+l^3[(B+A)(\frac{1}{3}-\frac{k}{4})-\frac{s}{3}]]$$

$$+I_{2,C_2z}+m_2r_2(r_2-x)+\frac{\pi\rho}{4}[\frac{sB^2}{4}(B+3A)[(\frac{k}{l})^3x^4+4(\frac{k}{l})^2x^3+2(\frac{k}{l})x^2-4x-lk^3-4lk^2-2lk+4l]$$

$$-s^2B(B+A)[-(\frac{k}{l})^2x^3+3(\frac{k}{l})x^2-3x+lk^2-3lk+3l]+\frac{s^3}{2}(3B+A)(\frac{k}{l}x-2x-lk+2l)+s^4(x-l)]]$$

(13)

In the same way are attained the relations for the **Rectangular hollow cross-section**. The differential equations of the elastic line for the upward motion in a vertical plane

$$w''(x) = -\frac{M_y(x)}{EJ_y(x)} \quad \text{with the boundary conditions} \quad w_{x=0} = 0 \;\wedge\; w'_{x=0} = 0 \quad (14)$$

and for the transverse movement in a horizontal plane

$$u''(x) = \frac{M_z(x)}{EJ_z(x)} \quad \text{with the boundary conditions} \quad u_{x=0} = 0 \;\wedge\; u'_{x=0} = 0 \quad (15)$$

by a variable cross-section cannot be integrated in a closed form. Therefore is used the numerical integration with values estimated from the documents of the robot Puma 560: $\ddot{\psi}=\ddot{\theta}=8$ rad/s^2, $m_2=11$ kg, $I_{2,C2,y}=I_{2,C2,z}=0.8$ kgm^2, $r_2=0.75$ m, $l=0.43$ m.

3. Optimization of the Arm Construction

3.1 Determination of the Optimal Height-to-Width Ratio

For equal vertical and horizontal rotational accelerations the bending moment M_y is greater than the bending moment M_z by the gravitational terms (last line in the Eq. (4)) according to Eq. (5). To achieve equal deflections in both directions symmetric closed hollow sections are to be used with a greater height than width.

<u>Example 1</u> Material: Steel

$\qquad$ E = 2.1×10^{11} N/m^2 $\qquad$ A = 3.84×10^{-3} m^2 $\qquad$ $\ddot{\psi} = \ddot{\theta} = 8$ rad/s^2

$\qquad$ ρ = 7850 kg/m^3 $\qquad$ s = 6 mm $\qquad$ k = 0

Fig. 2 shows the course of the deflections w and u at the end x=1 of the girder in vertical and horizontal direction for elliptic (index E) and rectangular (index R) hollow cross-sections. Under these conditions both cross-section forms show the same optimal ratio: v_{opt}^E (=A/B)=1.846, v_{opt}^R (=H/B)=1.837 at which both deflections are equal.

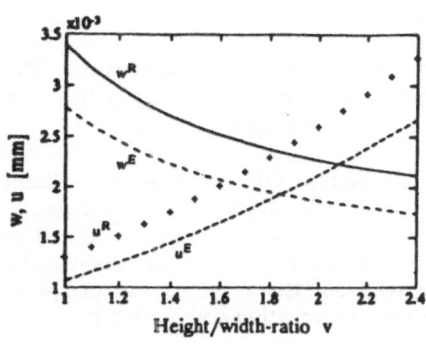

Fig. 2 Deflections $\qquad\qquad\qquad\qquad$ Fig. 3 Flexural rigidities

Besides, it is to be noted that by the same cross-sectional area (same mass) the elliptic hollow section has greater flexural rigidity than the rectangular hollow section.

<u>Example 2</u> Material: Aluminium

$\qquad$ E = 7.1×10^{10} N/m^2 $\qquad$ A = 3.84×10^{-3} m^2 $\qquad$ $\ddot{\psi} = \ddot{\theta} = 8$ rad/s^2

$\qquad$ ρ = 2700 kg/m^3 $\qquad$ s = 6 mm $\qquad$ k = 0

The optimal height/width-ratio depends from material (Young's modulus and density) too. Under the same geometric conditions and accelerations the aluminium girder has another optimal height/width-ratio for both section forms (v_{opt}^E = 1.793, v_{opt}^R = 1.787).

<u>Example 3</u> Material: Aluminium

$\qquad$ E = 7.1×10^{10} N/m^2 $\qquad$ A = 3.84×10^{-3} m^2 $\qquad$ $\ddot{\psi} = \ddot{\theta} = 4$ rad/s^2

$\qquad$ ρ = 2700 kg/m^3 $\qquad$ s = 6 mm $\qquad$ k = 0

Angular accelerations are important influence factor for the optimal height/width-ratio. For lower vertical and horizontal angular accelerations ($\ddot{\psi} = \ddot{\theta} = 4$ rad/s^2) the portion of the gravitational moment to the total bending moment is greater. In this case the optimal height/width-ratio for the girder is higher ($v_{opt}^E \doteq v_{opt}^R \doteq 2.39$).

<u>Example 4</u> Material: Aluminium

$\qquad$ E = 7.1×10^{10} N/m^2 $J_y^E = J_y^R = 1.33 \times 10^{-5}$ m^2 (=const.) $\qquad$ $\ddot{\psi} = \ddot{\theta} = 8$ rad/s^2

$\qquad$ ρ = 2700 kg/m^3 $\qquad$ s = 6 mm $\qquad$ k = 0

We assume that the rigidity of a girder with quadratical section for the upward motion in a vertical plane is determined for a certain allowed deflection of the link end. In this case it can be saved 11% of the mass by reduction of the width of the robot link for which the same deflection in the horizontal plane is achieved.

<u>Example 5</u> Material: Steel

$\qquad$ E = 2.1×10^{11} N/m^2 $\qquad$ $J_y^E = J_y^R$ $\qquad$ $\ddot{\psi} = \ddot{\theta} = 8$ rad/s^2

$\qquad$ ρ = 7850 kg/m^3 $\qquad$ s = 6 mm $\qquad$ k = 0

68

The cross-sectional area $A^E = 3.84 \times 10^{-3}$ m^2 of the elliptic hollow section is constant. Varying the height/width-ratio the dimensions of the rectangular section by constant thickness $s = 6$ mm are determined in such a way, that its areal moment of inertia J_y^R is equal to that of the elliptic section J_y^E. Then yields $J_z^E = J_z^R$ too. The cross-sectional area of the rectangular section (consequently the girder mass too) is greater than that of the elliptic section by equal areal moments of inertia. By nearly equal deflections the elliptic section has a smaller mass (by 6.2%).

3.2 Choice of the Favourable Rejuvenation Factor

The link mass decreases linearly with the increasing of the rejuvenation factor. The sequences of the deflections, however, become increasingly steep, specially from $k = 0.5 + 0.6$ upwards. Thus a rejuvenation factor $k = 0.5$ would be favourable.

An elliptic hollow steel link with the great axis $A = 100$ mm, thickness $s = 6$ mm, without rejuvenation ($k = 0$), with a mass $m = 9.486$ kg and an optimal height/width-ratio $v_{opt} = 1.815$ and under the same loading conditions of the example 1, has an end deflection $w = u = 4.77 \times 10^{-3}$ mm. A link with a rejuvenation factor $k = 0.5$ would need for the same end deflection w and u an $A_1 = 120$ mm (great half-axis at the fixation position), $s = 6$ mm, $v_{opt} = 1.78$ and $m = 8.562$ kg. Through this rejuvenation with $k = 0.5$ it can be saved in this case 9.75% of the mass.

4. Conclusions

The optimization of the robot arm construction begins with the determination of an optimal height/width-ratio of the hollow cross-section. Such an optimal ratio mainly depends on the angular accelerations and less on material, payload and rejuvenation factor. For angular accelerations 8 rad/s^2 (as by the robot Puma 560) the optimal height/width-ratio is 1.8. Introducing an optimal height/width-ratio by maintaining a given flexural rigidity for equal vertical and horizontal angular accelerations (8 rad/s^2) there can be saved 11% of the mass.

The dynamic comparison between the rectangular and elliptic hollow cross-section shows that the mass of the girder with rectangular cross sectional area by equal moments of inertia is greater than that of the elliptical one. For the same end deflection the elliptic section has a smaller mass by 6.2%.

5. References

Çaushi,I. (1993). Dynamische Einflüsse von Massen- und Steifigkeitsreduktionen auf die Bewegungsgrössen starrer und elastischer Industrieroboter. *Fortschritt-Berichte VDI* Reihe 18 Nr.120, VDI Verlag, Düsseldorf.

Desoyer,K., Kopacek,P. (1992). Industrieroboter: Stand und Entwicklungstendenzen. *Berichtband des 8. Österreichischen Automatisierungstages*. TU-Wien, 22. Oktober 1992.

Urbaneck,W. (1988). *Grundlegende Betrachtungen zur Konstruktion und Auswahl von Materialien für den Bau von Leichtbaurobotern ausreichender Steifigkeit*. Diploma-Thesis, Technical University of Vienna.

Walser,A. (1991). *Grundlegende Analysen für Armkonstruktionen von Leichtrobotern ausreichender Steifigkeit*. Diploma-Thesis, Technical University of Vienna.

Design and Construction of a Modular Robot with Eulerian Joints

M. M. Gola, D. Botto
Mechanical Department
University of Technology, Turin, Italy

Abstract: The aim of our research is the construction of a robot arm with non conventional architecture for applications in non-industrial environments. In this work we develop a machine for agricultural applications, such as fruit picking. The arm must be hollow and enable the fruit to be conveyed through it, thus saving cycle time. In this paper we analyze an original kinematical configuration particularly suited to this purpose, and develop a preliminary design. At last we present the final design and discuss architectural and constructional details.

1 Kinematics of the joint

The adopted kinematics configuration is characterized by a number of consecutive hinges, each of them formed by two equatorial surfaces in relative rotation about a common axis, such axis forming an angle to the arm axis.

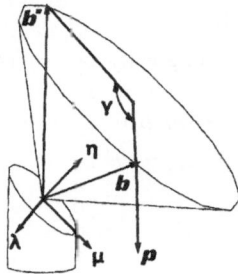 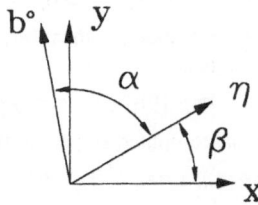

(Fig. 1)

In fig. 1, axes (λ, μ) define such equatorial couple of planes, and the following arm has an axis $\vec{b}$ at an angle α to the normal axis $\vec{\eta}$ about which the rotation of the planes takes place. Axis $\vec{b}$ is thus provided of a conical motion. The generic rotation of the arm from an initial position b^0 about the axis $\vec{\eta}$ is represented by the matrix $R_{\gamma,\vec{\eta}}$. So the next arm location after a rotation γ is given by $\vec{b} = R_{\gamma,\vec{\eta}} \vec{b^0}$ where

$$
R_{\gamma,\vec{\eta}} = \begin{bmatrix}
\eta_x^2 (1 - \cos\gamma) + \cos\gamma & \eta_x \eta_y (1 - \cos\gamma) - \eta_z \sin\gamma & \eta_x \eta_z (1 - \cos\gamma) + \eta_y \sin\gamma \\
\eta_x \eta_y (1 - \cos\gamma) + \eta_z \sin\gamma & \eta_y^2 (1 - \cos\gamma) + \cos\gamma & \eta_y \eta_z (1 - \cos\gamma) - \eta_x \sin\gamma \\
\eta_x \eta_z (1 - \cos\gamma) - \eta_y \sin\gamma & \eta_y \eta_z (1 - \cos\gamma) + \eta_x \sin\gamma & \eta_z^2 (1 - \cos\gamma) + \cos\gamma
\end{bmatrix}
$$

We choose our reference frame so that the versor $\vec{\eta}$ is located in the plane xy. The position of this versor is then definite by the angle β measured from the x axis. We also choose a zero rotation, $\gamma = 0$, when the arm is on the plane xy. Then the vectors components are:

$$\eta_x = \cos\beta, \quad \eta_y = 0, \quad \eta_z = \sin\beta \quad ; \quad b_x^0 = 0, \quad b_y^0 = b\cos(\alpha + \beta), \quad b_z^0 = b\sin(\alpha + \beta)$$

where b is the arm length. In the study of the individual hinge, a mass is placed at the end of $\vec{b}$. The computation of the motor torque which is necessary to withstand the external load is performed through the equation of virtual work. The virtual work of the external forces due the gravity only is $\delta L_e = \vec{P}^T \cdot \vec{\delta b} = m\,\vec{g}\cdot d\,R_{\gamma,\vec{\eta}}/d\,\gamma\,\vec{b}\,\delta\gamma$, while the virtual work of the internal forces, namely the motor torque, is $\delta L_i = \vec{C_M}\cdot\vec{\eta}\,\delta\gamma$. Since $\delta L_i = \delta L_e$ the motor torque necessary to balance the forces due to gravity is:

$$C_M = m\,g\,[\,(\,\eta_x\,\eta_z\,\sin\gamma - \eta_y\,\cos\gamma\,)\,b_x + (\,\eta_y\,\eta_z\,\sin\gamma + \eta_x\,\cos\gamma\,)\,b_y + (\,\eta_z^2 - 1\,)\,\sin\gamma\,b_x]$$

In our design $\alpha = 45°$ and we have two possible cases:

a)   If the hinge is the first one, it is constrained to the ground, and the angle β is positioned at $45°$. It is easily found that the maximum motor torque necessary to balance the external load is requested when $\gamma = 90°$ ($\pm 180°$) and its value is $C_{MAX} = m\,g\,b\,/\,2$

b)   If the hinge is one of the following ones, the angle β can assume any value, so we deduce the maximum torque from the diagram showed in the figure 2. This diagram shows the torque C_M necessary to balance the force of gravity in all the range of the variable β and γ. We see that the maximum torque is reached when $\beta = 0°\,\pm 180°$ and $\gamma = 90°$ ($\pm 180°$). In this case the torque is $C_{MAX} = \sqrt{2}\,m\,g\,b\,/\,2$

An evident advantage of this type of hinge consists in the fact that only a part of the external load is supported directly by the motor. The force of gravity produces bending moments at the hinge: this is calculated after introduction of the versors $\vec{\lambda}$ and $\vec{\mu}$. This versors together the versor $\vec{\eta}$ form a frame of reference and their components are:

$$\lambda_x = 1, \quad \lambda_y = 0, \quad \lambda_z = 0 \;\; ; \;\; \mu_x = 0, \quad \mu_y = \sin\beta, \quad \mu_z = -\cos\beta$$

Figure 1 shows the global and local reference frame. The force of gravity moment computed respect to the $\vec{\lambda}$ axis is:

$$M_{\vec{\lambda}} = \vec{b}\times\vec{P}\cdot\vec{\lambda} = -m\,g\,[(\,\eta_y^2\,(\,1 - \cos\gamma\,) + \cos\gamma\,)\,b_y^0 + \eta_y\,\eta_z\,(\,1 - \cos\gamma\,)\,b_z^0]$$

while the moment computed respect to the $\vec{\mu}$ axis is:

$$M_{\vec{\mu}} = \vec{b}\times\vec{P}\cdot\vec{\mu} = m\,g\,(\,-\eta_z\,b_y^0 + \eta_y\,b_z^0\,)\,\sin\gamma\,\sin\beta$$

The total moment is then $M = \sqrt{M_{\vec{\lambda}}^2 + M_{\vec{\mu}}^2}$.

Figure 3 shows the total moment as a function of angles β and γ. Using this diagram we see the point of maximum load on the support, and refer to it for the choice of the bearings. Using a couple of a such hinges, tilted between them, we obtain the eulerian joint. The figure 4 shows an arm formed by two joints and four hinges. In the first picture the joints have no rotation. In the second one only the hinges of the upper joint have been rotated while in the third picture all the hinges have been rotated.

2 Preliminary design

The preliminary design has been developed following the steps below:

A)   the arm dimension, its length and its inner diameter, the rotation axis slope, and the maximum external load are chosen; in the preliminary design we take into account only static load; moreover we use a simplified model where we do not take into account the offset in the hinge;

B)   the bevel gear pair dimension are chosen; they are restrained by the hinge geometry and the motor position; moreover the inner and outer radius of the bevel gear must be chosen so that it can enter in the oil case;

C)   the motor and the reducer are chosen and the performance are estimated;

D)   we compute the tooth stress and the Hertzian bearing stress; if limit values are exceeded, go back to the step B;

E)   at last we compute the bearings, the shaft and the linkage stress; if limit values are exceeded, go back to the step A.

All this operations are automated and are accomplished by a software. Table 1 shows the components chosen for the first joint.

Motor			
Kind	Torque	Speed	Mass
Moog D313-L15	1.6 [Nm]	6000 [rpm]	1.8 [Kg]
Reducer			
Kind	Max output torque	Max input speed	Mass
Harmonic drive HIUC-32	304.0 [Nm]	7000 [rpm]	3.2 [Kg]
Inner bearing : Kaydon KC090XP0 Outer bearing : Kaydon KC110XP0			
Bevel gear : module 3, width of tooth 10 mm, ratio 3.73			
	Peach diameter	Angle	Teeth
Gear	291 [mm]	71 [deg]	97
Pinion	78 [mm]	19 [deg]	26

(Tab. 1)

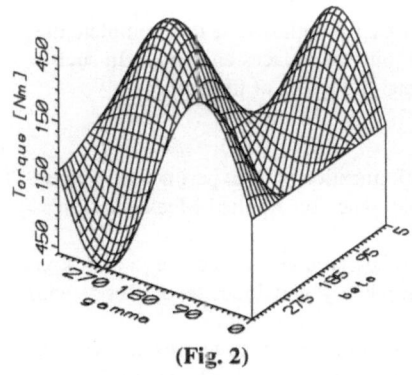

(Fig. 2)

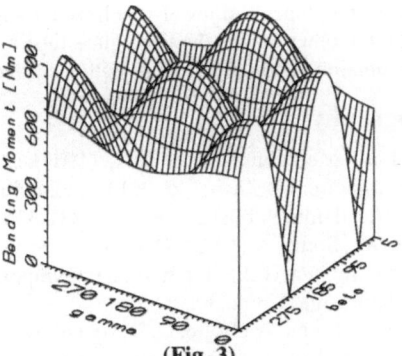

(Fig. 3)

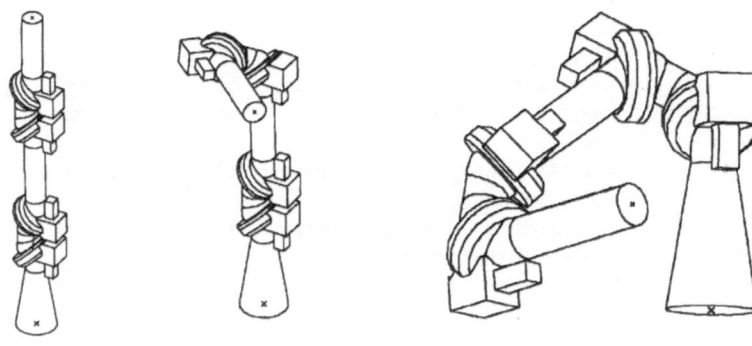

(Fig. 4)

3 Joint description

Figure 5 shows the final drawing of a typical joint. The eulerian joint has been realized using the elbowed part 1. The ends of this part are circular plate tilted of 45 degrees respect to the local $\vec{y}$ axis. The circular plates form an angle of 90 degrees between themselves. The bevel gear 2 is integral with the elbowed part, and their linkage form the housing for the race of the thin section bearings 3. The upper and the lower parts, both 4, support the motors 5 and the harmonic drive reducers 6. The output power is transmitted to the bevel pinion 7 through the shaft 8. When the motors are operated, part 4 rotates about axis $\vec{\eta}$. To take up slack in the bevel gear a pair of ring nuts 9 is used. A second ring nut is used as an anti-loosening device; a threaded section on the shaft accommodates the nuts.

The upper and lower parts 4 enclose gear 2 in the sealed case 10. The ends of such sealed case together with the rings 11 form the second housing for the thin section bearings 3. Oil-tightness is guarantied by the grommets 12. In further versions it may be possible to use directly sealed bearings. Obviously the bevel gear size are restrained by the bearing dimension. In order to easily assemble the harmonic drive 6 the pinion 7 and the shaft 8 in box 13, we chose to drill the lower side of the box 13. The cap 14 close this hole and form the seat for the shaft. In this way we increase the stiffness of the whole group.

The end of the upper and lower part have a flange to link a fibre glass tube that complete the joint. On the other end of the fibre glass tube a second joint take places and so on. In such a way we obtain a modular structure with a desired number of degrees of freedom.

Bibliography

Gola, M. M., Mebrahtu, Y., Soma, A,. (1989). Design optimization of an experimentally tuned robotic structure. In: Zheng, Z. (Ed.). International Conference on Applied Mechanic, International Academic Publishers, Beijing, 331-336

Gola, M. M., Soma, A.(1992). Design of a carbon fibre robot: architectural choices and design balancing. In: n/a (Ed.). International Symposium on Theory and Practice of Robot and Manipulators, n/a, Udine, n/a

Gola, M. M., Soma, A., Botto, D. Design of two non conventional robot structures. In: Nemec, B. (Ed.). First International Meeting on Robotics in Alpe Adria Region, Slovenian Robotic Society, Portoroz, 91-98

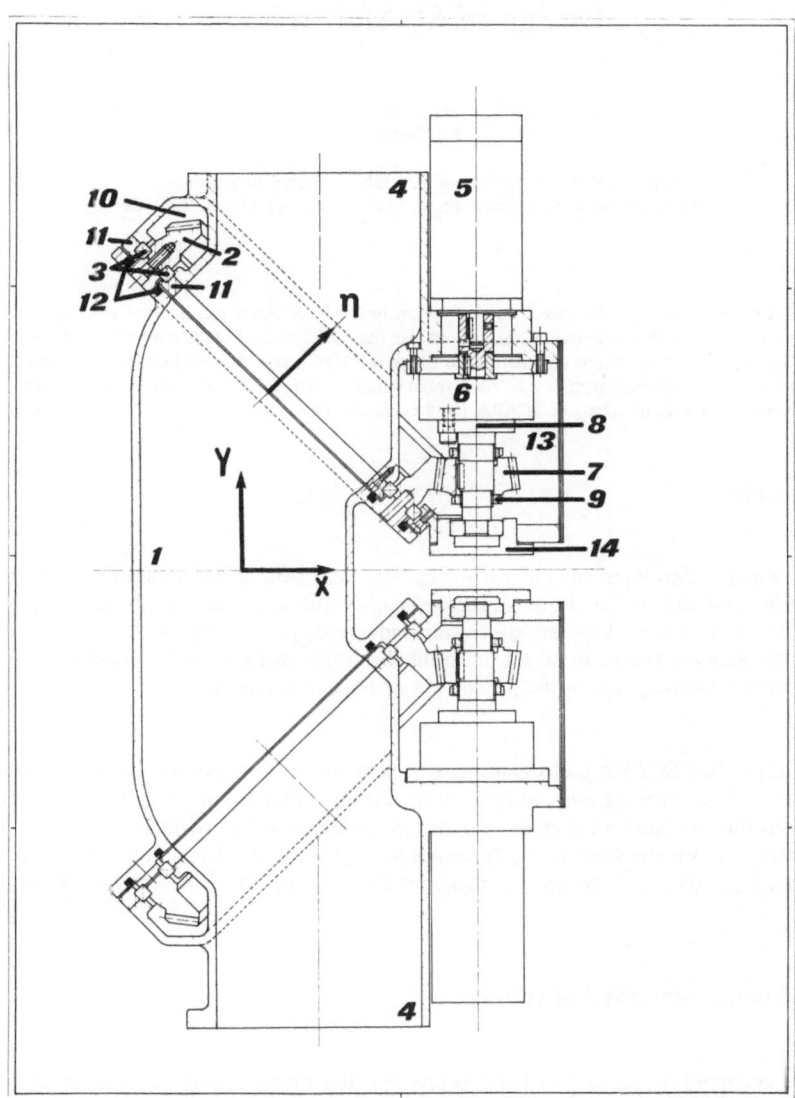

(Fig. 5)

This research has been developed within the frame of " Progetto Finalizzato Robotica" of C.N.R of Italy, grant number 92.01069.67.

Design of SCARA robot

D. Noe
M. Skubic

Laboratory for Handling and Assembly Systems Automation,
Faculty of Mechanical Engineering, University of Ljubljana, Slovenia

ABSTRACT: In the first part the paper presents the analysis of a SCARA manipulator working area. The second part concerns with the dynamic analysis on the basis of Lagrange equation. The moments in the particular link and the actual torque courses related to the real time and real robot form was calculated. Both analyses are part of the development of a SCARA robot named AVRO for a Slovenian enterprise. The results will be introduced in a future study of SCARA robot characteristics.

1 Introduction

SCARA robots, developed in the early eighties, represent a breakthrough in assembly automation (Makino, 1980). Robots' performance, large working area in the x - y plane, suitable height in z axis, low sensitivity for load speed, positioning and control accuracy, enabled the SCARA robots to be set up in the electronic printed circuits assembly, also in the automotive industry and in the production of home appliances.

When studying the SCARA performances, two items were kept in mind. First we wanted to analyze the robots working area, to find out the maximal dimensions of pallets that could be placed into this area and analyze the parameters influencing the working area dimensions. The second item was the study of the dynamics for a planned SCARA robot, with positioning accuracy of ± 0.05 mm, capable for handling the loads up to 12 kg, to find the suitable motors.

2 Initial considerations and results

A simple assembly robot, such as the two-link SCARA robot, can be analyzed using simple trigonometry, rather than homogeneous transformation equations. The position of the point B (x, y) for the two-link robot (Fig. 1) is determined by a vector equation:

$$^{\circ}P_{0,2} = {}^{\circ}P_{0,1} + {}^{\circ}P_{1,2} \tag{1}$$

where $p_{i,j}$ is the vector from the origin of frame i to the origin of frame j as seen from the reference frame.

Using simple trigonometry, we can find the x and y components of the vector $p_{0,1}$ and $p_{1,2}$. When these values are substituted into equation (1) we obtain the geometric model of the robot.

$$^{\circ}P_{0,1} = (l_1 \cos(\theta_1) \, , \, l_1 \sin(\theta_1)) \tag{2}$$

$$^{\circ}P_{1,2} = (l_2 \cos(\theta_1 + \theta_2) \, , \, l_2 \sin(\theta_1 + \theta_2)) \tag{3}$$

$$^{\circ}P_{0,2} = \begin{bmatrix} p_x \\ p_y \end{bmatrix} = \begin{bmatrix} x \\ y \end{bmatrix} = \begin{bmatrix} l_1 \cos(\theta_1) \, , \, l_1 \sin(\theta_1) \\ l_2 \cos(\theta_1 + \theta_2) \, , \, l_2 \sin(\theta_1 + \theta_2) \end{bmatrix} \tag{4}$$

The orientation of the first vector is θ_1, and the second one, is the sum of the joint angles ($\theta_1 + \theta_2$).

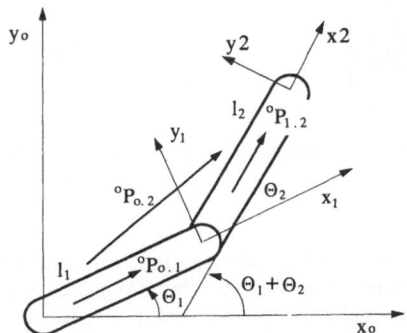

Fig. 1. SCARA type manipulator in zero position.

The working area of a two-link robot, and the dimensions of the pallet can be determined by knowing the vector to the point B_2 at robot's maximal reach ($p_{0,2}$). Here both links are in a straight line ($\theta_2 = 0$). This value is called r_2. The vector to the point B_2 represents the minimal reach of the robot, and this value is called minimal reach radius (r_1). The value of r_1, at given maximal radius r_2, is determined by the angle between links θ_2 and by the ratio of the lengths of the links (Fig. 2).

$$r_1 = \sqrt{l_1^2 + l_2^2 + 2 \cdot l_1 \cdot l_2 \cdot \cos\theta_2} \tag{5}$$

$$r_2 = l_1 + l_2 \tag{6}$$

The dimensions of the largest pallet that could be placed into this area are defined by the following equation:

$$(b+r_1)^2 + a^2 - r_2^2 = 0 \tag{7}$$

To find the suitable configuration of the two-link robot for a given working area, or for a given pallet, one has to determine the length of both links and the maximal turning angle of the second link. From the equation it can be deduced that at a given maximal turning angle of the second link, we get the maximal working area when both links are equally long. If the maximal turning angle $\theta_2 = 135°$ we get $r_1 = 0.763 \, l_1$, and when $l_1 = 2l_2$ the r_1 is 27,8 % larger than at equally long links.

76

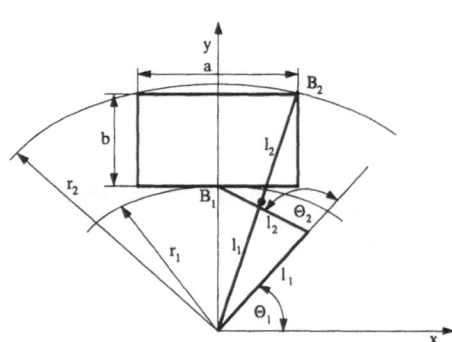

Fig. 2. The SCARA robot working area.

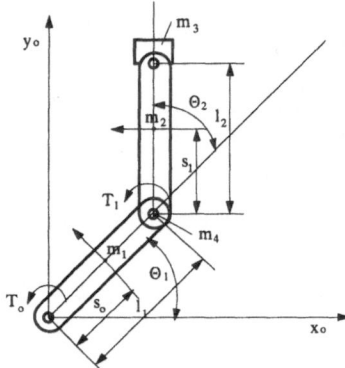

Fig. 3. Parameters of two-link SCARA
type manipulator.

Besides, when developing new robots, the algorithm for maximal working area determination can also be used for choosing a suitable robot from a database. In the robot's data, the lengths of both links (l_1 and l_2) and the maximal turning angles θ_1 and θ_2 are given. To find a suitable robot for a given working area it is necessary to determine the vectors to the points B_1 and B_2 (Fig.2).

Dynamics deals with the motion of bodies. This includes kinematics, which is the study of the motion of bodies without reference to the forces that cause the motion, and kinetics, which relates these forces to the resulting motions. So, dynamics is the branch of mechanics that deals with the motion of bodies under the action of forces. On basis of this definition, the dynamics behaviour of a robot is described in terms of the time rate of changes of the arm configuration to the torques by its actuators.

There are two problems, the problem of inverse dynamics and the problem of forward (direct) dynamics. If a vector of joint position, velocities, and acceleration is given, and the required vector of joint torques is calculated, we get the inverse dynamics problem. It is possible to describe this problem by the following functional connection:

$$^{R}T_{n} = f(\ ^{R}\theta_{i}\ ,\ ^{R}\omega_{j}\ ,\ ^{R}\alpha_{k}\ ,\ robot\ configuration\ ) \tag{8}$$

The problem of forward dynamics: a vector of applied joint torque is given, and the resultant manipulator motions have to be calculated. The problem of forward dynamics is described by the following functional connection:

$$^{R}\alpha_{h} = g(\ ^{R}\theta_{i}\ ,\ ^{R}\omega_{j}\ ,\ ^{R}T_{n}\ ,\ robot\ \ configuration\ ) \tag{9}$$

For dynamics analysis, a two-link manipulator is placed into the x - y plane. The lengths of links are designated by l_1 and l_2, the corresponding masses are m_1 and m_2, and the motion trajectories are defined by the angles θ_1 and θ_2.

77

Forces and torques affecting the robot's links can be calculated by the Lagrange equations. The Lagrangean formulations describe the dynamic behaviour of a robot in terms of work done by, and energy stored in the system. The robot is treated as a black box that has an energy balance. The generalised Lagrange equation when applying n independent generalised coordinates $q_1 \ldots q_n$ is:

$$L(\ddot{q}_n, q_n) = K - U \tag{10}$$

where L is Lagrangian function, K is kinetic energy and U is potential energy. For SCARA the following is true: $U=0$ and $q_n = \theta_i$ and the equation for the joint torque are:

$$^RT_i = \frac{d}{dt}\left(\frac{\partial K_i}{\partial \omega_i}\right) - \frac{\partial K_i}{\partial \theta_i} \tag{11}$$

For choosing both motors for SCARA - AVRO robot the kinetic energies for both arms, second motor and the load were calculated. When implementing these values into the Lagrange equation we get the torques in the first link:

$$T_0 = D_{11}\alpha_1 + D_{12}\alpha_2 + D_{122}\omega_2^2 + D_{11}\omega_1\omega_2 \tag{12}$$

Where T_0 is torque in the first link, D_{11} is effective inertia torque part of the second arm, D_{12} is effective inertia of the second arm related to the first one, D_{122} is centripetal force part, $D_{11}\,\omega_1\,\omega_2$ is the Coriolis force.

And for the second link:

$$T_1 = D_{21}\alpha_1 + D_{22}\alpha_2 + D_{22}\omega_1^2 \tag{13}$$

Here T_1 is torque in the first link, D_{22} is effective inertia torque art of the second arm, D_{21} effective inertia of the first arm related to the second one and D_{221} is part of centripetal force.

The actual torque, which the motors have to manage is therefore:

$$M_i(t) = J_{im} + J_{ig} = \alpha_i(t) \cdot N_i + \frac{T_i(t)}{N_i} \cdot n_i \tag{14}$$

Here J_{im} is the inertia torque of the i-th motor, J_{ig} is inertia torque of the i-th gear, N is the transmission ratio of the gear, and η_g is efficiency of the gear.

The actual torques for both motors were calculated numerically by a suitable computer program. When calculating, a trapezoid course of velocities was supposed, and maximal angle was given at $\theta_1 = \theta_2 = 4.71$ rad. The transmission ratio for gears and maximal revolutions for motors were also given. The travelling time from one extreme point to another was given by 1.5 s. These are heavy working conditions and the motors have to manage heavy loads.

The plot of ideal velocity course is shown in Fig. 4a, and the actual velocity course in Fig.4b. The differences between ideal and actual courses are obvious.

78

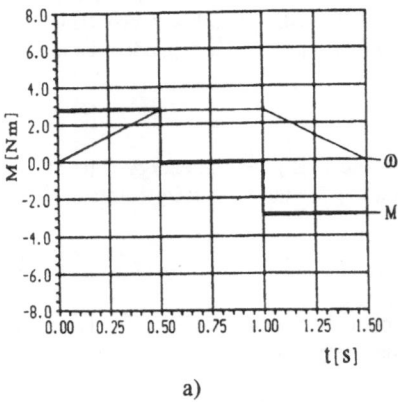
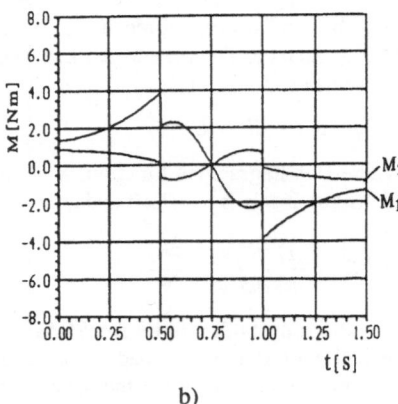

a)   b)

Fig. 4. Ideal (a) and actual courses (b) of torque.

3 Conclusions

Two problems of constructing and/or choosing a SCARA robot were studied. First the working area of the robot was analyzed. The results of the analysis can be used for the determination of the lengths of the arms also for choosing a suitable robot for a given assembly task.

For the purpose of robots construction and determination of motors, the torques were analyzed using the Lagrange equations. A special computer program for calculation of torques at a given configuration and moving time, and trapezoid shape of velocity course was developed.

4 References

Makino H., N. Furura: Selective Compliance Assembly Robot Arm, Proceedings of the 1st International Conference on Assembly Automation, Brighton 1980, P. 77 to 86.

McKerrow,P. J.: Introduction to Robotics, Addison-Wesley Publishing Company Sydney 1991, P. 347 to 398.

Design Measures for Lightweight Industrial Robots

I.D. Çaushi
Institute for Handling Devices and Robotics
Technical University of Vienna
and
Institute for Mechanical Constructions
Polytechnical University of Tirana

Abstract. In this paper the effects of mass reduction by industrial robots are investigated with the aim to obtain optimized lightweight structures of industrial robots. Formulation of mathematical robot models and simulations are performed by a software (Autolev), which makes use of Kane's method. The robot system IGM Limat RT 280-6 is modelled as a rigid multibody system with 6 DOFs. All six electromagnetic controlling effects of driving units, inertia and gyroscopic effects of rotors, friction moments and payload are included. Effects of all these system parameters which will be important for future design of appropriate controllers of lightweight robots are examined. By analyzing effects of different mass reductions an energy criterion to optimize the mechanical structure of lightweight industrial robots is formulated.

1. Introduction

Worldwide are being carried out several multidisciplinous scientific works aiming at designing lightweight robots being concentrated on the three following directions: 1. mechanical construction and component improvements (Desoyer and Kopacek, 1992), 2. modeling and simulations (Çaushi, 1993), and 3. appropriate control design.

Efforts to obtain high positioning accuracy and repeatability have brought about massive structures of industrial robots having a very low degree of material utilization and which cannot move at the required speeds. Therefore, the first step in perfecting the mechanical system of industrial robots should consist in improving their technical level.

2. Modeling

As specimen for the simulation investigations is used an industrial robot with six revolving rigid links, which is destined for welding. Modeling and symbolic generation of the dynamic equations is made with the software package for dynamic analysis of multibody systems AUTOLEV based on the Kane's method.

In the most studies of robot dynamics, a reduced order model from the point of view of the basic degrees of freedom is used. Since for the dynamics of lightweight robots, which have to move fast and accurately, each element could have a considerable influence, this study aims at creating a comprehensive model. This modeling implicates six local coordinate frames for the links and an additional coordinate system for the payload, to associate the inertia moments to the reflection of the payload mass. Six other coordinate frames are related to the driving rotors, whose whole effects can be considered thereby. As to dynamic friction effects, they are modelled through a constant, a velocity proportional and an exponential component with coefficients gained through measurements (Çaushi, 1992). The driving electric DC motors are modelled each as a first order lag unit with a subsequent integrator (Çaushi, 1992). The six nominal voltage values of the DC motors are used as input quantities for the model.

3. Analyzing Simulation Results

3.1 Influence of Rotors, Friction and Payload

Internal rotating driving components - like rotors or gears - have considerable effects on the robot dynamics. The velocity of the robot endpoint v_l is an important parameter of motion (especially for welding robots) which is influenced from the inertia and gyroscopic effects of the rotors.

Through the model without rotor terms results at the end of 0.5s a velocity of the robot endpoint 22.5% lower than the real velocity (calculated with the exact model). The acceleration of the robot endpoint (important for certain payloads) is influenced by the rotors too: The acceleration peak value of the robot endpoint is feigned by the model without rotor terms by 30% higher. Because of this are overvalued the inertial forces too. That can be of importance for the dimensioning of a lightweight robot.

Although the rotors have small inertia parameters, their kinetic energy is 32 per cent of that of the links. This considerable portion is due to the high rotation speeds of the rotors (high gear ratios).

Friction causes an indolent dynamic behaviour. Under the influence of friction the angles of rotation $q_i(t)$ and angular velocities $u_i(t)$ ($i=1,...,6$) get smaller values (up to -25% at the sixth link) at the end of the time interval in consideration. For the model with friction after the acceleration phase (after the torque peak values) greater driving torques $T_{ai}(t)$ ($i=1,...,6$) to overcome the friction torques are needed.

The influence of a payload 7.5 kg on the motion of the endpoint of the robot (resulting from angles of rotations of the links, see Tab. 1) is considerable.

The effects of neglecting the rotors, friction and the influences of a payload 7.5 kg on the quantities $Ta_i(t)$ (motor torques), $a_i(t)$, $u_i(t)$, $q_i(t)$ (angular accelerations, velocities and angles of rotation), ($i=1,...,6$), a_L (acceleration of the robot endpoint) and $E_{el,m}$ ($m=1,...,6$) (electric energy requirements) are putted together on the Tab. 1.

Each of the model components (rotors, friction or payload) being included in the model causes a diminution of the angles of rotation, angular acceleration peak values of the links, acceleration peak value of the robot endpoint and increase of the power peak values or energy requirements of the motors. From the percentages of the effects of these model components it can be drawn the conclusion that by dynamic simulations and in order to develop highly accurate control algorithms for lightweight industrial robots the driving rotors, friction effects and payload are to be included into the robot model.

3.2 Analyzing Energy of Motion of Each Robot Link

During the supposed motion (all six motors supplied with their respective nominal voltages) the robot links gain different kinetic and potential energy, according to their mass, velocity and the achieved position. By means of the developed program for the dynamic simulation of the studied robot Limat RT the curves of the kinetic and potential energy of each link are obtained (see Fig. 1). This energy has to be brought up from the driving motors.

Tab. 1 *The effects of neglecting rotors, friction and payload influences [%]*

Influence	Rotors	Friction	Payload
$\Delta T_{a1/\text{peak value}}$	-3.83	-1.64	5.05
$\Delta T_{a2/\text{peak value}}$	-2.98	-2.39	8.77
$\Delta T_{a3/\text{peak value}}$	-28.36	-3.73	14.51
$\Delta T_{a4/\text{peak value}}$	-50.85	-0.85	1.69
$\Delta T_{a5/\text{peak value}}$	-1.49	-1.48	105.00
$\Delta T_{a6/\text{peak value}}$	-84.58	-0.50	-
$\Delta a_{1/\text{peak value}}$	10.66	1.93	-14.03
$\Delta a_{2/\text{peak value}}$	28.89	5.71	-52.94
$\Delta a_{3/\text{peak value}}$	33.70	3.23	-34.68
$\Delta a_{4/\text{peak value}}$	152.01	5.36	-19.54
$\Delta a_{5/\text{peak value}}$	-14.51	3.71	-44.01
$\Delta a_{6/\text{peak value}}$	218.42	15.93	-6.20
$\Delta u_{1/T=0.5s}$	-	3.24	5.74
$\Delta u_{2/T=0.5s}$	-	7.60	2.36
$\Delta u_{3/T=0.5s}$	-	4.85	1.51
$\Delta u_{4/T=0.5s}$	0.15	5.31	-29.46
$\Delta u_{5/T=0.5s}$	0.12	3.83	-33.87
$\Delta u_{6/T=0.5s}$	0.23	24.56	-2.30
$\Delta q_{1/T=0.5s}$	3.17	3.81	-0.63
$\Delta q_{2/T=0.5s}$	4.75	8.13	-19.67
$\Delta q_{3/T=0.5s}$	3.20	4.80	-13.66
$\Delta q_{4/T=0.5s}$	15.62	5.47	-20.01
$\Delta q_{5/T=0.5s}$	1.89	3.77	-46.68
$\Delta q_{6/T=0.5s}$	93.33	16.08	3.45
$\Delta a_{1/\text{peak value}}$	0.30	0.05	-0.67
$\Delta E_{el,1}$	-14.37	2.61	15.96
$\Delta E_{el,2}$	-8.87	-20.97	29.25
$\Delta E_{el,3}$	-22.39	-28.26	88.23
$\Delta E_{el,4}$	-41.93	-14.84	23.39
$\Delta E_{el,5}$	-10.77	-25.64	323.40
$\Delta E_{el,6}$	-73.68	-12.28	-12.56
$\Delta E_{el,\text{total}}$	-18.39	-16.17	45.24

The largest part of the energy of motion is consumed by the second, third and fourth link (Fig. 1). This fact leads to the conclusion that the masses of these links have to be reduced to a higher extent to minimize the energy consumption.

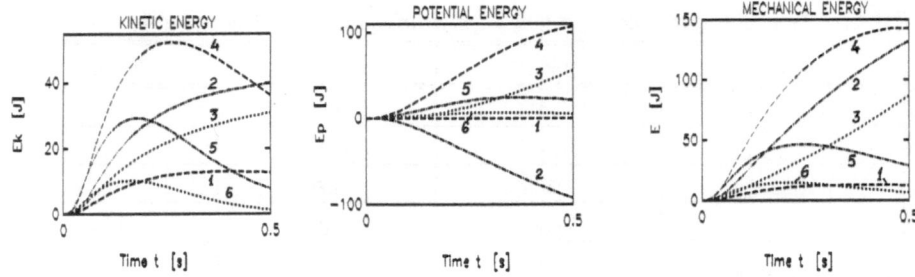

Fig. 1 *Kinetic and potential energy of the links.*

3.3 Effects of Mass Reductions

Using special construction design as integrated framework (Feyerabend, 1988) and/or lightweight materials etc. (Desoyer and Kopacek, 1992), the mass of the robot links can be considerably reduced keeping stiffness relatively high.

In the first series of investigations is realized a dynamic simulation with a mass reduction by 30% of all links (model B). The robot links retain their size, i.e. their radii too. The inertia moments of the links are also reduced by 30% too. As it turned out from the analysis of the energy of motion of the links, the greatest energy portions are needed for the motion of the 2nd, 3rd and 4th link. Therefore it is expedient for a design optimization (with energy consumption as index) to reduce at first the inertia parameters of these three links. Accordingly in a second case there are investigated the dynamic influences of a 40% mass reduction of the selected (2nd, 3rd and 4th) links (model C).

Tab. 2 *Influences of mass reductions on E_{kin} and E_{pot} of the robot links [%]*

i	1	2	3	4	5	6
$\Delta E_{kin,i}$ (B)	-31.82	-27.94	-31.43	-32.93	-32.22	-30.0
(C)	-3.18	-37.50	-40.57	-41.46	-	-
$\Delta E_{pot,i}$ (B)	-	-25.81	-21.05	-28.10	-33.33	-32.3
(C)	-	-36.13	-34.21	-38.02	-	-

In the developed simulation program the electric energy consumption (see Tab. 3) is calculated in the case of driving and braking through electric motors, and holding in standstill through mechanical brakes (Çaushi, 1991).

The advantage of the mass reduction of selected links by model (C) is the fact, that, in spite of a greater total mass as by model (B), the model (C) requires less energy of motion. In Tab. 3 can be seen that by means of only 10% more reduction of the links 2 and 3 it is more energy to be saved as by means of the 30% reduction of the (less moved) link 1. This effect is yet enlarged because the first link doesn't move in the vertical direction and its potential energy remains constant (Fig. 1).

To drive the mass reduced models smaller driving torques are needed. Therefore there can be chosen the next smaller motors (at least the first three). Not only through diminution of motor masses but also through that of inertia moments of the rotors can be saved more energy. Through the reduction of the link masses (including motor masses too) in each driving unit can be saved energy (Tab. 3).

Tab. 3 *Influences of mass reduction on electric energy requirements [%]*

Motor	m = 1	2	3	4	5	6	Σ_m
$\Delta E_{el,m}$ (B)	-24.12	-22.39	-10.92	-8.46	-16.9	-0.1	-18.5
(C)	-24.5	-25.69	- 5.02	-1.54	+ 1.56	-1.7	-18.5

In Tab. 3 is to be seen that by model (C) with 40% reduction of the links 2, 3 and 4 (which corresponds to a 26% total reduction) the same energy amount (18.5%) as by model (B) with 30% equally distributed total reduction can be saved.

4. Conclusions

Dynamic simulations are performed on a developed compute model including all six links (6 DOF), friction effects, payload, inertia and gyroscopic effects of rotors and electromagnetic control effects of the driving units. They show that it is important for the design of controllers for lightweight industrial robots to include all these elements into the model of the robot system.

Industrial robots have to be mechanically improved through optimized mass distribution according to criteria of minimal energy consumption. In this article are shown which effect different cases of weight reduction of the links for a welding industrial robot would cause on the energy consumption. The selection of the links whose weight is to be reduced more than the weight of the others is oriented towards analysis of the energy of motion (kinetic and potential) for each link.

For an energy optimization of robot constructions the inertia parameters of the links are to be progressively diminished with increasing distance from the base, so that during the motion by nominal motor voltages the courses of the quantity $E_{kin} + |E_{pot}|$ of the links as much as possible near each other lie.

5. References

Çaushi, I.D. (1992). Investigations on leightweight industrial robots. In: *Robot Control 1991 (Syroco '91). Selected papers from the 3rd IFAC/IFIP/IMACS Symposium, Vienna, Austria, 16-18 Sept. 1991*. Pergamon Press, Oxford New-York Seoul Tokyo, pp. 537-541.

Çaushi, I.D. (1993). Dynamische Einflüsse von Massen- und Steifigkeitsreduktionen auf die Bewegungsgrössen starrer und elastischer Industrieroboter. *Fortschritt-Berichte VDI* Reihe 18 Nr. 120, VDI Verlag, Düsseldorf.

Desoyer, K., Kopacek, P. (1992). Industrieroboter: Stand und Entwicklungstendenzen. *Berichtband des 8. Österreichischen Automatisierungstages*. TU-Wien, 22. Oktober 1992.

Feyerabend, F. (1988). Systematic optimization of a robot arm structure. *The Industrial Robot*, 15(4), 219-22.

Designing Manipulators for a Robotized Manufacturing

M. Ceccarelli, L. Carrino
Department of Industrial Engineering
University, Cassino, Italy

Abstract: The mechanical design of a proper manipulator can be recognized of fundamental importance to achieve a rational system layout for a robotized manufacturing. In the paper, an attempt is illustrated, both from a teaching viewpoint and with design purposes, to formulate a design methodology for manipulators taking into account the specific characteristics of a robotized manufacturing with advanced composite materials. A general analytical formulation is proposed by means of an integration between manufacturing and design concepts to stress the problem requirements and constraints, and to suggest future developments.

INTRODUCTION

The need of industrial competitiveness requires the use of new materials and the development of economic production processes. Consequently, the automation development, occurred for the mass production, is needed by means of very flexible systems even in manufacturing of few pieces with advanced composite materials. In fact, the needs of high precision in manufacturing and great reliability in the final product characteristics require an automation, at the most, of the manipulative operations. Moreover, since the production with advanced composite materials is usually limited to few pieces, a robotized manufacturing seems to be appropriate to achieve a high flexibility for products with certain differences in size and shape.

In this context, robots can be considered as fundamental devices for the manufacturing automation since a robotized system flexibility depends strongly from the manipulator versatility. Consequently, a manipulator design must be achieved taking also into account the peculiarities of the production environment, (Dorf, 1984). Nevertheless, a mechanical design of a manipulator is usually obtained by considering a suitable limited description of the manufacturing task requirements from a kinematic (Lenarcic et. al., 1989) or mechanical viewpoint, (Seering, 1985). Although it is recognized that the design parameters are closely related to the task, (Schraft, 1985), and system design considerations can be provided, (Engelberg, 1977), (Vertut, 1981), an analytical formulation of the manipulator design problem is usually approached with a synthetic indexing of the manipulator characteristics, that may provide only an idea of the manufacturing task which has been designed for.

ON THE DESIGN OF A ROBOT

A robot is a complex system whose components, which are systems as well, can be recognized in: a) the mechanical structure, usually referred as the manipulator, which interacts with the environment to carry out the physical task; b) the actuators, which provide the robot motion and the system functionality; c) the control and power unit, which supply the necessary energy and its regulation; d) the sensory equipment, providing sensorial capabilities which are needed to know the robot state and the environment nature; e) the informatic facilities (software and hardware), which supply the computational power and the Artificial Intelligence capabilities.

Since the beginning of the Robot Technology, the design was conceived with the aim to achieve universal or multi-task robots, (Engelberg, 1977), (Vertut, 1981). Nevertheless, several industrial applications do not require usually high versatility level in a sense that certain manipulative tasks can be limited to a priori determined operations. Versatility can be related to the mechanical possibilities of a manipulator to achieve several mechanical tasks with respect to manipulative requirements. Synthetically, this can be also considered through a number of degrees of freedom. On the other hand, flexibility of a robot is usually referred to the capabilities of reprogramming the task within the possibilities of the manipulator. From these considerations, the mechanical possibilities of a manipulator may be properly designed with respect to determined manipulative tasks, which nevertheless can be accomplished with a proper robot programming.

Moreover, a robot becomes obsolete very quickly and its payback has been stretching, in the last years, to one-year in some case. Therefore, economic requirements and constraints must be also considered from a technical viewpoint as fundamental in a robot design.

A ROBOTIZED MANUFACTURING WITH ADVANCED COMPOSITE MATERIALS

The increasing replacement of metal parts with advanced polymeric composites is largely justified by remarkable improvements in structural effectiveness. This goal is obtained through the most known properties of advanced composite materials: high strength and stiffness, the chance to define mechanical and physical characteristics depending on the structural application. The fatigue behaviour, environment strength and impact strength are further characteristics to prefer advanced polymeric composites to metallic materials, (Schwartz 1987). Nevertheless, products of advanced composite materials show a very high final cost, which is due at the most to the manufacturing process and to the row material costs. Particularly, a 60-70% of the product cost can be attributable to the manufacturing technology, since it still requires a great amount of manual labour, (Crivelli Visconti et.al. 1989). A further major problem consists in the need of absence of defects, which turns out in product quality in terms of repeatability and reliability of the product properties.

One of the most adopted manufacturing technologies also for complex 3-D products is the winding. A product is made by reinforcement strands by means of hand winding on a specific tool with poor repeatability and reliability, (Rolston 1982). In fact, the final characteristics of a product may depend strongly from the filament posture which is obtained through a filament winding characterized, basically, by the path, the twisting, the orientation and the tension in the roving operation. Therefore, it is thought that improvements on product characteristics and in the manufacturing cost can be achieved mostly by an automation of the winding operation.

Moreover, since actually the advanced composite materials are not used for mass production but at most for high performance characteristics, the production is usually limited in the number of pieces, which eventually can be repeated with different size and small shape changes. Therefore, in some case it can be convenient to prefer the automation by means of suitable robotic manipulators which can perform determined manipulative task of a roving filament for a priori selected product family.

In this paper the main functional and system requirements in a robotized manufacturing with advance composite materials has been related with the winding filament manipulation.

A DESIGN FORMULATION OF MANIPULATORS FOR A ROBOTIZED MANUFACTURING

From the above mentioned reasoning, it seems convenient to adopt a design philosophy by which a robot may be designed for a specified task, (Seering, 1985), so that a robotized manufacturing system can be selected and optimized from the beginning in all its components. This requires that the interactions among the robot components might be taken into account from the first step of conceiving the manipulator structure, Fig.1. The common practice of adjusting with suitable equipment and components an existing mechanical structure can produce dystrophic robots, which often are huge with respect to the actual task. Basic concept for a specified task design is to understand the mutual influences of the system components and to formulate performance indices which can evaluate general characteristics, (Schraft 1985), as well as peculiarities of the specified application.

In Fig.1 a flow chart for general design features is illustrated by stressing the fact that the mechanical design is strongly affected by mechanical requirements and, moreover, by the design features and requirements of the robot components which are thought to be necessary to meet the application goals. The difference between the synthesis of a kinematic chain and the mechanical design of a mechanical structure has been emphasized since it is thought of primary importance to have a robot whose kinematic characteristics are well proportioned with respect to a specified task application and the component equipment. The particular emphasis on the kinematic chain of manipulators has been stressed in a previous paper, (Ceccarelli et.al., 1992), in which the synthesis problem has been focalized from a general kinematic viewpoint. It has been also recognized the importance of kinematic chain characteristics, as the workspace, in an optimized layout of a manufacturing cell, (Hoang, 1992).

This paper is an attempt to propose, on the basis of the previous considerations, a design methodology of the mechanical structures of robots for specified task applications in manufacturing with advanced composite materials according to the flow charts of Fig.1 and, particularly, Fig.2. Fig.2 stresses the prominence of the manufacturing task on the manipulator design in a sense that specific manipulating characteristics require suitable manipulator structures, both from a kinematic and mechanical design viewpoints.

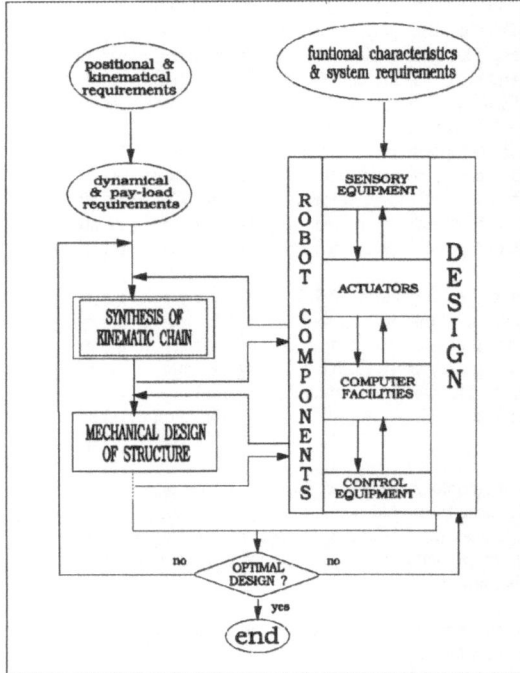

Fig.1. A general design procedure for robot systems.

manipulator. The objective functions should be considered as fundamental characteristics to be optimized and they should be typical both of manipulator structures and of the robotized manufacturing. In addition, the constraint functions can be thought as requirements to be met both for the structural design and the automation layout. From a purely kinematic viewpoint the manipulator's design parameters are usually considered the link parameters of the manipulator chain, (Ceccarelli et.al., 1992).

A fundamental problem consists in formulating properly f and g from integrated considerations between design and manufacturing so that the design of mechanical structures according to the block of Fig.2 may give a proper basis for an optimal design of the robot system. To achieve this goal, it is thought that the requirements and the characteristics of the industrial application and the overall system should yet considered in the kinematical synthesis of robots. This requires that a first idea of the robotized manufacturing solution should be conceived in some extent since the selection of the manipulator kinematic chain.

Then, an indexing formulation of the individuated characteristics and requirements is needed in a form to be included in the optimization problem.

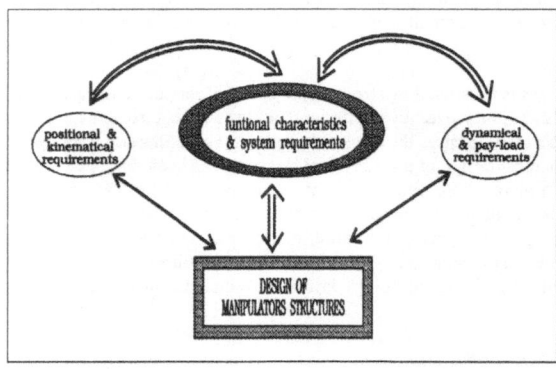

Of course, this formulation activities require a strong interaction between the designers and the technologists. Moreover, this modelling activity involves back and forward thinking on the system conceivement. Further-more, this can produce a re-designing even in the manufacturing process. An attempt in this engineering integration can be described in formulating some functional characteristics and manufacturing requirements to include in an explicit form of the integrated formulation of the optimization problem, according to the design outlines of Fig.2.

Fig.2. Integrated design requirements for manipulators.

Particularly, the problem of designing manipulators for advanced composite materials can be formulated through suitable objective and constraint functions deducible both from peculiarities of a specific manufacturing and manipulator design considerations. The design parameters in this case of study have been considered the time independent parameters of the manipulator chain: the link length a_i, the link offset d_i and the twisting angle α_{i-1} for all the n links.

Thus, it has been thought useful to consider some fundamental characteristics through the following formulation of objective functions:
- the manufacturing mean cycle time T_W, expressing a time interval which is necessary to complete the winding operation for products of a typical family

$$f_1 = T_W \tag{1}$$

This can be thought greatly affected by the manipulative capabilities of the manipulator, which can evaluated synthetically by means of workspace and dexterity evaluation. T_W can be computed through a path simulation for the winding operation to be considered within the design process;
- the manipulator encumbrance, which can be expressed through the manipulator length as

$$f_2 = \sqrt{\sum_{i=1}^{n} \left(a_i^2 + d_i^2 \right)} \tag{2}$$

- the manipulator degrees of freedom, expressing the number of n revolute joints and the links in a manipulator structure, in the form

$$f_3 = n \tag{3}$$

as a measure of manipulator versatility;
- the cost of the robotized manufacturing through the payback period for the manipulator, calculated through the expression

$$f_4 = \frac{C + I}{H(L + M - R) - K_V \Delta v H L + D} \tag{4}$$

where C is the cost of the robot and its tooling, I is an installation cost, H is the number of working hours per month, L is the hourly cost of labour, M is an hourly cost of saved materials and products, R is the cost of running and maintaining of the robotized manufacturing, $K_V \Delta v$ is an hourly cost of increasing or decreasing the productivity for a change in the winding velocity per piece; D is a yearly cost of the depreciation. In some extent this cost evaluation can be thought related to the manipulator size by considering, for example, that both C and R may increase when n or the total length increase.

In addition, constraints can be deduced by considering:
- workspace limits and precision points in the form

$$X_j \leq X_j^* \qquad j=1,..,J \tag{5}$$

where the J vectors X* represent given workspace coordinates, within or acrossing which the manipulator may work;
- the winding filament torsion φ within a limit angle φ_w of rotation of the strand given as

$$\varphi \leq \varphi_w \tag{6}$$

and an orientation error $\Delta\theta$ between a strand axis and the design axis for the product within given bounds that are dictated by fiber requirements in the form

$$\Delta\theta_{min} \leq \Delta\theta \leq \Delta\theta_{max} \tag{7}$$

which both involve a suitable orientational capability of the manipulator during the manufacturing;
- a posture velocity of the winding filament which is constrained within

$$v_{min} \leq v_w \leq v_{max} \tag{8}$$

where the lower limit may be imposed by the working time before the natural polymerisation of the material and at most by productivity rate; the upper bound can be dictated by strength and glueyness characteristics of the material.

Finally, it is also convenient from a practical engineering point of view to consider some solution domains for the design parameters in the form

$$x_{min} \leq x \leq x_{max} \tag{9}$$

Eqs.(1)-(9) express an analytical formulation of a design problem of a manipulator for a robotized manufacturing according to Fig.2.

CONCLUSION

In this first study, it has been thought convenient to formulate the design problem of a robotized manufacturing in a general analytical form, suitable for future extension and developments. A design problem has been formulated referring to a specific manufacturing with advanced composite materials. Thus, the optimization problem has been proposed through a specific formulation of objective and constraint functions taking into account some peculiarities both of the manipulator design and the manufacturing with advanced composite materials. Some general considerations have been developed to illustrate a design methodology which considers, from the beginning, the manipulative task which a robot is designed for.

REFERENCES

Ceccarelli, M., Cuadrado, J.I., Mata, V.A. (1992). Design of Kinematic Chains for Robotic Manipulators. Journal Proyecto 2000, Pulsar, Barcelona, n.74, 67-77.

Crivelli Visconti, I., Langella, A., Carrino, L., Fernandes, C., Montesarchio, B., (1989). Design and Fabrication of a Composite Gear Structural Element. In: Proceedings of International Conference on Composite Material, Milan.

Dorf, R.C. (1984). Robotics and Automated Manufacturing. Reston Publ. Co., Reston.

Engelberger, J.F. (1977). Designing Robots for Industrial Environments. Mechanism and Machine Theory, 403-412.

Hoang, K., Fenton, R.G. (1992). Determination of Robot's Location in Manufacturing Cell. In: Proceedings of 23rd Int.Symposium on Industrial Robots. Barcelona,121-125.

Lenarcic, J., Stanic, U., Oblak, P. (1989). Some Kinematic Considerations for the Design of Robot Manipulators. Jnl. Robotics & Computer-Integrated Manufacturing, 235-241.

Rolston, J.A. (1982). Filament Winding. In: Composite Design Guide Vol.3:Processing and Fabrication Technology. University of Delaware.

Scheinman, V., Roth, B. (1985). On the Optimal Selection and Placement of Manipulators. In: Proceedings of the Fifth CISM-IFToMM Symposium on Theory and Practice of Robots and Manipulators. Kogan Page, London, 39-45.

Schraft, R.D., Wanner, M.C. (1985). Determination of Important Design Parameters for Industrial Robots from the Application Point of View: Survey Paper. In: Proceedings of the Fifth CISM-IFToMM Symposium on Theory and Practice of Robots and Manipulators, Kogan Page, London, 423-429.

Schwartz, M.M. (1987). Composite Materials Handbook. McGraw Hill, New York.

Seering, W.P., Scheinmann, V. (1985). Mechanical Design of an Industrial Robot. In: Nof, S.Y. (Ed.). Handbook of Industrial Robotics, Wiley, New York, 29-43.

Vertut, J., Liègeois, A. (1981). A General Design Criteria for Manipulators. Mechanism and Machine Theory, 65-70.

Optimal Stochastic Design of a Parallel Robot for Tolerance

P. B. Zobel, P. Di Stefano
Dipartimento di Energetica
Universita' di L'Aquila, Aquila, Italy

Abstract: This paper presents both a sensitivity analysis to the dimensional errors and a statistical analysis of the effects of manufacturing tolerances for the Delta robot. Delta is a four degrees of freedom (dof) robot with a parallel structure. The kinematic model of the robot has been adjusted to consider geometrical and assembly parameters. Mechanical errors are analysed for the 3-Sigma band of confidence level through a stochastic model of the robot. The working area in which the manufacturing errors sensitivity has the minimum value is presented and a method to choose the optimal design tolerances is also suggested.

1 Introduction

In the robotics field the positioning accuracy of the end-effector of the robot is one of the main characteristics. The kinematic parameters affect the positioning accuracy, because they present errors due to the machining of the geometric links and to the assembly. To obtain a good accuracy these errors have to be limited by choosing proper dimensional tolerances.

A certain number [2-6] of papers have been presented in the last years on closed loop mechanisms using a stochastic model with the goal of minimizing the structural and the mechanical error in the design step.

In the present paper a stochastic model of a parallel robot named DELTA is proposed. This robot was developed by R. Clavel [7,8] at the Ecole Polytechnique Federale de Lausanne for pick and place operations of light weight objects. For these operations a very fast robot, with low inertia and high rigidity of structure, is required. This robot has 4 d.o.f., a parallel structure which gives a high rigidity and very light links.

At first we modified the direct kinematic model proposed by Clavel. This was necessary because the kinematic models are based on the symmetry of the structure, while the sthocastic model is characterized by an asymmetric structure. Then a sensitivity analisys was performed on the positioning accuracy of the end-effector, due to the dimensional errors in the links and assembly of the robot. Kinematic parameters of the robot was considered as random variables for a 3σ band of confidence level and a methodology for the evaluation of the optimal working volume for positioning error is proposed.

Finally a parameter that links the positioning accuracy to the manufacturing cost is defined and aimed at an optimal allocation of design tolerances. The results obtained in a planar trajectory of the robot are presented in curves for two values of links tolerance.

2 The Statistical Approach

Due to the tolerance of the link lengths and the assembly dimensions, the end-effector position of DELTA robot deviates from the nominal position. In a statistical model of the robot this deviation can be analyzed for a 3σ band of confidence level which corrispond to a 99.73% probability of success. The random output P, i.e. the end-effector position of the robot, in a stochastic model depends on the random variables z_i, i.e. links length of the robot:

$$P = P(z_1, z_2, \dots, z_n) = P(Z) \tag{1}$$

The 3σ band of confidence level of output P is defined as 'positioning error' $e_p = 6\sigma_p$.
We can expand (1) in Taylor series about the mean value of each random variables and negletting the second order and higher terms, the mean and the variance are obtained as:

$$E(P(Z)) = P(E(Z)) \qquad VAR(P(Z)) = \sigma_P^2 = \sum_{j=1}^{n}\left(\frac{\partial P}{\partial z_j}\right)^2 \sigma_{z_j}^2 \qquad (j=1,2,...,n) \qquad (2)$$

The partial derivatives of equation (2) are referred to as sensitivities (s_i). The random variables, considered in this paper, are the random link lengths LA_i and LB_i, and the random assembly dimensions RR_i and θ_i for $\{i=1,2,3\}$ and n=12. These twelve chosen variables are the most important characteristic variables of the DELTA robot. The same methodology can be used including other geometric parameters of the robot. The sensitivities of the end-effector position with respect to the dimensional parameter $(s_j = \partial P / \partial z_j)$ are obtained as a numerical derivation by the kinematic direct model modified by the authors.
Each z_i is a random variable and is characterized by a mean E and a variance value VAR,

$$E(z_i) = z_{in} \qquad VAR(z_i) = \sigma_{z_i}^2 = \left(\frac{t_{z_i}}{6}\right)^2 \qquad (i=1,2,3) \qquad (3)$$

where z_{in} is the nominal link length or assembly dimension and t_{zi} is the tolerance width on the z_i dimension. Then the tolerance value for each parameter considered is $t_{z_i} = 6\sigma_{z_i}$ and the positioning error e_p in the three space directions of the DELTA's absolute frame is:

$$e_{p_k}^2 = (6\sigma_{P_k})^2 = \sum_{j=1}^{n}\left(\frac{\partial P_k}{\partial z_j}t_{z_j}\right)^2 \qquad (k=x,y,z) \qquad (4)$$

3 Manufacturing cost and optimal design

The positioning error of the robot has a minumum value at lower bounds of tolerances while the manufacturing cost has a minimum value at the upper bounds. With the aim of optimizing these two contrasting objectives many techniques were proposed[11,12].
In this research a method to choose optimal manufacturing tolerances, based on a specific parameter which takes into account both manufacturing cost and the positioning quality, was proposed. The simplified relationship between the cost C_a and the tolerances is $C_a = K_a / t_a$ (a=1,2.....n) where t_a is the size of the tolerance of the ath dimension and K_a is a constant. In order to obtain the trend of costs with respect to the tolerance it is useful to derive the previous relationship for the ath dimension:

$$\frac{\partial C_a}{\partial t_a} = -\frac{K_a}{t_a^2} \qquad (5)$$

The positioning error e_a for the generic tolerance t_a is written as $e_a = s_a t_a$. Now deriving respect to t_a and dividing by equation (5) the relation between e_a and the cost C_a is obtained:

$$\frac{\partial e_a}{\partial C_a} = -\frac{s_a t_a^2}{K_a} \qquad (6)$$

Using a Pareto approach this parameter, evaluated for each dimensional characteristic, suggests the best choice of tolerances to reach a lower position error with the least cost.

4 Numerical simulations and discussion

The DELTA robot consists basically of three parallel kinematic chains closed at the end-effector, figure 1. They are arranged at 120° to each other and each chain is actuated by an electric motor fixed to the superior base. This structure has three d.o.f. and the end-effector is a travelling plate that conserve an orientation parallel to the superior base. The motion of the end-effector is the result of the movement of the three robot arms, mounted on the base through three couples of parallel rods. These parallel rods ensure that the plate always remains parallel to the base during motion. The actuators are mounted on the superior fixed base and the only moving masses consist of the kinematic structure so it is possible to have a very high speed.

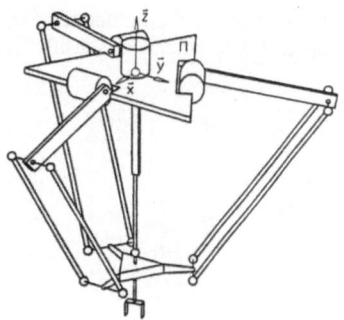

Fig.1 The DELTA robot

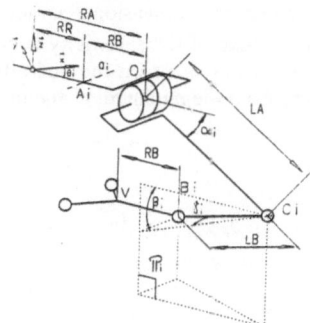

Fig.2 Characteristic parameters of the robot

The numerical simulations have been carried out for the 1988 ISIR DELTA version. In table 1 the characteristic nominal dimensions of this version and the tolerance values used for the calculation of the positioning error are shown.

If the three dimensions LA_i have the same manufacturing tolerances, it is possible to consider all of them once calculating the statistical sensitivity parameter $s_{LA} = \sqrt{(\Sigma s^2_{LAi})}$ so that the statistical interference of the three random variables is considered.

	arm 1	arm 2	arm3	tolerance
LA	0.2 m	0.2 m	0.2 m	0.032 mm
LB	0.38 m	0.38 m	0.38 m	0.049 mm
RA	0.19 m	0.19 m	0.19 m	0.029 mm
RB	0.025 m	0.025 m	0.025 m	-
θ	0.0°	120°	240°	0.014°

Table 1

Each element LB consists of two links which are joined together by ball and socket articulations at each of their extremities. In this paper each of the elements LB are considered as one dimension.

In figure 3 the isosensitivity curves for the statistical sensitivity parameters, related to all the dimensional characteristics analized, are presented in a square working plane parallel to the superior base(z=-0.3 m and side=0.36 m).

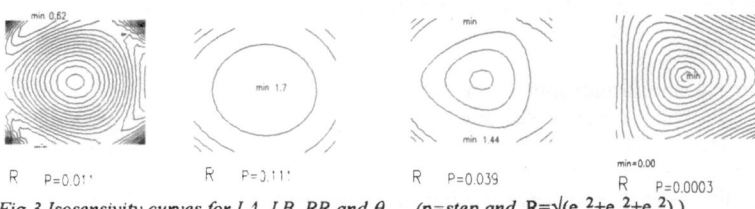

Fig.3 Isosensivity curves for LA, LB, RR and θ   (p=step and $R=\sqrt{(e_x^2+e_y^2+e_z^2)}$)

It is interesting to observe that the sensitivity values are often higher than unity, this means that the effect of the input tolerance is amplified. The sensitivity values for the θ parameter, as shown in the figure, seem to be very little. It is important to observe that it is not correct to compare a dimensional parameter (s_θ=[mm/deg]) with dimensionless ones.

The range of the sensitivity is quite wide in the working plane; infact the influence of the dimensional error on the end-effector position changes up to 60%, when the sensitivities vary from the minimum to the maximum value.

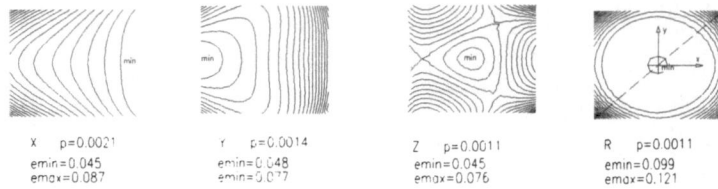

Fig. 4 Iso-error curves and trajectory studied in the x-y plane

The positional error for each dimensional characteristic (e_{LA}, e_{LB}, e_{RR} and e_θ.) is composed to evaluate the end-effector positioning error so that the statistical interference of the twelve random variables is considered.

In figure 4 the result is shown considering the tolerance values shown in table 1. Then a rectilinear trajectory of the end-effector, on the working plane examined (z=-0.3 m) for the same tolerance values, was analized.

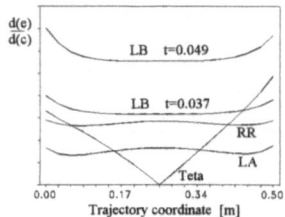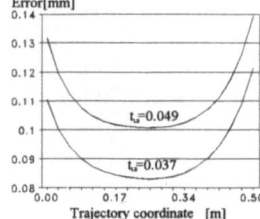

Fig.5 Absolute value of d(e)/d(c) and positioning error for the trajectory studied (K_{LA}=1.00, K_{LB}=1.53, K_{RR}=0.90, K_θ=0.42)

On this trajectory the parameter (6) was evaluated for each dimensional characteristic (fig.5). This figure shows that LB presents the highest value of this parameter for the whole trajectory. So that a riduction of the tolerance value for LB gives the best results of positioning error for a certain manufacturing cost. Then we reduced this tolerance value for

LB from t = 0.049 mm to t = 0.037 mm, this value was chosen to reach the parameter values of the other characteristics. The figure 5 shows the positioning error for these two values of LB's tolerance.

5 Conclusions

In this paper a method for the stochastic design of the DELTA parallel robot has been presented. This methodology has been applied considering four geometrical characteristics, allowing to choose the optimal tolerances. The optimal working area, that exibits the minimum sensitivity to dimensional errors, has been also evaluated by this method. It is the basis to design the operative layout of the robot for pick and place operation. An example of the optimal tolerance choice, in the case of a rectilinear trajectory, to reach a better positioning accuracy at the least cost, is presented. This example shows that with a 76% reduction of the tolerance of the arms LB, a reduction of up to 84% of the positioning error is obtained. The continuation of this study forsees the use of this method to define the optimal dimensions of the robot's links that minimize the sensitivity values, i.e. the positioning error.

REFERENCES

[1] Broderick P.L. and Cipria P.J. (1988). "A Method for Determining and Correcting Robot position and Orientation Errors Due to Manufacturing". ASME Journal of Mechanisms, Transmissions, and Automation in Design, Vol.110, 3, pp.3-10

[2] Dhande S G. and Chakraborty J. (1973). "Analysis and Syntesis of Mechanical Error in Linkages - A Stochastic approach". ASME Journal of Engineering for Industry, Vol.95, 8

[3] Lee S.J. and Gilmore B.J. (1991). "The Determination of the Probabilistic Properties of Velocities and Accelerations in Kinematic Chains With Uncertainty". ASME Journal of Mechanical Design, Vol.113, 3, pp.84-90

[4] Lee W.J. and Woo T.C. (1989). "Tolerance Volume Due to Joint Variable Errors in Robots". ASME Journal of Mechanisms, Transmissions, and Automation in Design, Vol.111, 12, pp.597-604

[5] Bhatti P.K. and Rao S.S. (1988). "Reliability Analysis of Robot Manipulators". ASME Journal of Mechanisms, Transmissions, and Automation in Design, Journal of Mechanisms, Transmissions, and Automation in Design, Vol.110, 6, pp.175-181

[6] Fenton R.G., Cleghorn W.L. and Fu Jing-fan (1989). "Allocation of Dimensional Tolerances for Multiple Loop Planar Mechanisms". ASME Journal of Mechanisms, Transmissions, and Automation in Design, Vol.111, 12, pp.465-470

[7] Clavel R. (1988) "DELTA, a fast robot with parallel geometry". Proc.Int.Symposium on Industrial Robots, pp.91-100.

[8] Clavel R. (1991) "Conception d'un robot parallèle rapide a 4 degres de liberte". PhD dissertation (THESE n° 925), Ecole Polytecanique Fèdèrale de Lausanne, Switzerland.

[9] Clavel R, Stevens B.S. and Rey L. (1992) "The Delta Parallel Structured robot, yet more Performant thtough Direct Drive". Procs of 23rd International Symposium on Industrial Robots, pp.485-493

[10] Beomonte Zobel P. (1991) "Study of Positioning Accuracy of Robot DELTA". Internal Report - Ecole Polytechnique Fèdèrale de Lausanne, Switzerland.

[11] Rhyu J.H. and Kwak B.M. (1988). "Optimal Stochastic Design of Four-Bar Mechanisms for Tolerance and Clearance". ASME Journal of Mechanisms, Transmissions, and Automation in Design, Vol.110, 9, pp.255-262.

[12] Drozda T.J. and Wick C. (1983) "Tool and Manufacturing Engineers Handbook - vol. I - Machining". SME, New York.

USIS - An advanced 3D-Robot Simulation System

L. Bauer, R. Stetter, C. Woenckhaus
Institute for Tool Machines and Industrial Engineering
Technical University, Munich

1 Abstract: The number of robot installations in industry has increased significantly over the past several years. To support planning of new complex robot applications, several offline programming and simulation systems have been developed.
The Institute for Tool Machines and Industrial Engineering (iwb) of the Technical University of Munich is developing its own offline programming and simulation system, called USIS (Universal Simulation System) (Wrba, 1990, Woenckhaus et al., 1993). The goal of the development is to create a system which, on the one hand, contains a precise model description of the work cell to simulate robot behaviour accurately. On the other hand, new programming and planning techniques will be implemented to reduce programming efforts significantly. This paper presents some results from this research work.

2 USIS in General

A basic feature of USIS is the design of a robot work cell. The user is able to select different cell components from several libraries, for example about 30 different robot types, containing up to six different control languages. The geometrical description of parts, which are not stored in this libraries can be passed over from CAD-systems using different interfaces, like VDAFS, IGES or SLA. The components have to be placed interactively after being chosen. This task is done by the use of several interactive functions to change the locations of components. In the next step programs are generated offline in the robot specific control language. USIS offers different interactive functions for graphical robot programming.

The application of USIS is not limited to robot simulation, also the programming and simulation of tool machines (Schrüfer, 1992) and the simulation of manual work systems (Kummetsteiner, 1992) is feasible.

3 Advanced Model Description

In order to get usable simulation results exact model descriptions are necessary. This does not only affect the geometrical model description but the real component behaviour. Some examples for this extended model description are physical effects, like gravity, flexibility or friction, which are implemented into the USIS (Stetter, 1992). E.g. under the effect of simulated gravity a part, which is released from a gripper, is forced to fall down. Coupled with fast algorithms for collision detection it is possible to calculate the final position of impact.

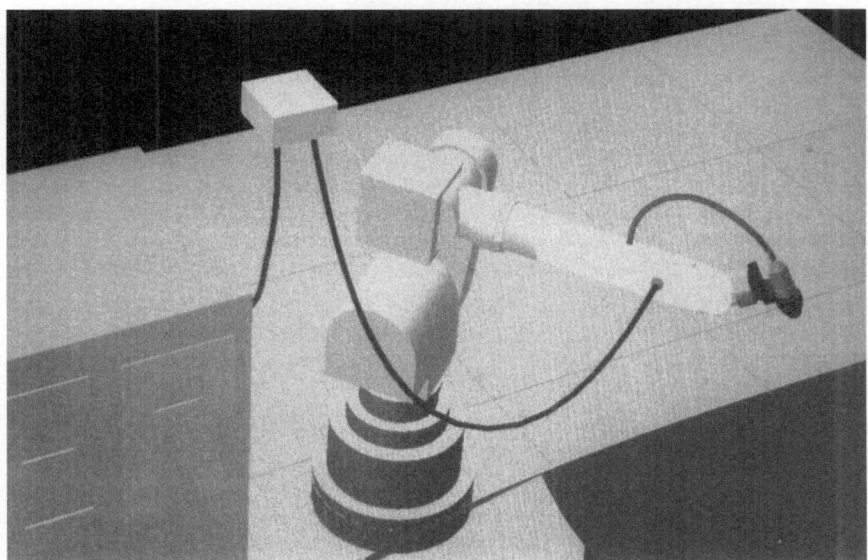

Fig. 1: Welding Robot with External Supply Hoses

With the development of faster and lighter manipulators robot dynamics becomes increasingly important. The determination and solution of the system of nonlinear differential equations is done by integrated dynamic simulation packages (Woenckhaus, 1990). Another topic is the simulation of flexible material behaviour, for example a flexible pipe between the robot and the gripper (Fig. 1). In addition to this different sensor systems, like a laser sensor (Milberg and Schuster, 1989), a laser scanner or a CCD-camera (Stetter, 1993) are integrated into the USIS system, too. Fig. 2 shows a flexible manufacturing cell with a robot, which is supplied with a CCD-camera in the gripper. The actual view area of this camera is visualized in a separate window.

4 Graphical Robot Programming

An advantage of offline programming is the possibility to develop robot programs without using the real robot. The most common method of graphic robot programming is the graphic teach-in. To generate a whole path the user has to identify each certain robot positions. This method is very time consuming and does not use all available information, which the simulation model contains.

A more advanced method of graphical robot programming is realized by using the so called frame mode. Positions are no longer stored in absolute coordinate values but with respect to a reference coordinate system linked to a component. If this component is placed to another position during the planning process, the program positions are adjusted automatically.

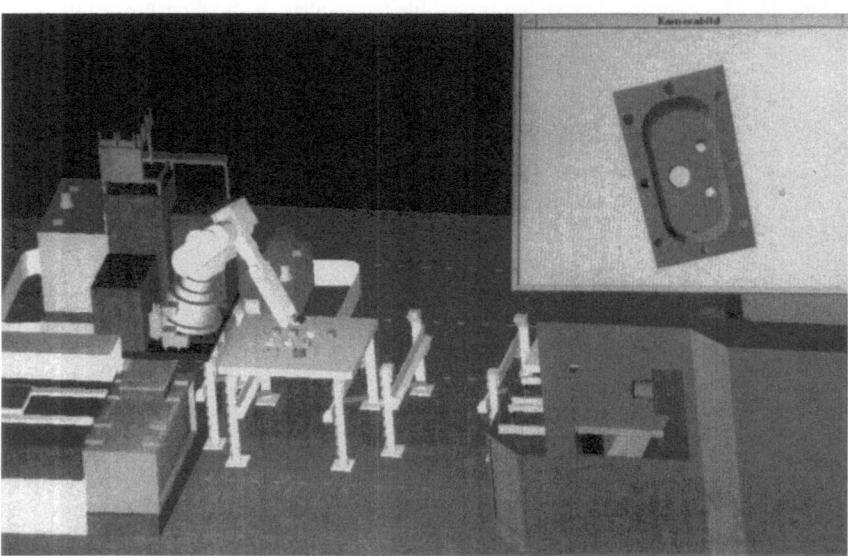

Fig. 2: Simulation of a CCD-Camera

The most comfortable method of offline robot programming is the automatical path planning. In USIS this is done by a self developed algorithm, which searches a collision free path between an user defined starting point and a destination point (Stetter, 1990) (Fig. 3). To find this robot track, it is necessary to have exact informations about the assembly cell environment. These informations are obtained from several sensor systems, which are also implemented in USIS. The result of path planning is a complete optimized robot program, which is written in the specific robot control language. Fig. 3 snapshots one point during the path planning process.

In addition to the path planning algorithm grasp operations are planned automatically by the system, too. The planning process is done in two steps. First an optimized arrangement between gripper and part is choosen with respect to geometric restrictions. Then the robot program is automatically generated for the complete grasp sequence. Path and grasp planning can also be used in combination. So the planner only has to identify the part to grasp and the start and destination location. USIS then generates the necessary program statements for a collision free grasp sequence, the transfer and the droping motion.

5 Numerical Optimization

There is still a great potential to optimize the component arrangement in robot work cell. To minimize planning efforts the three-dimensional simulation system is linked to a numerical optimization package (Woenckhaus, 1992). An optimization algorithm will suggest changing locations of different cell components. After this the robot programs will be automatically updated to the new locations using the so called frame mode. Subsequently this layout

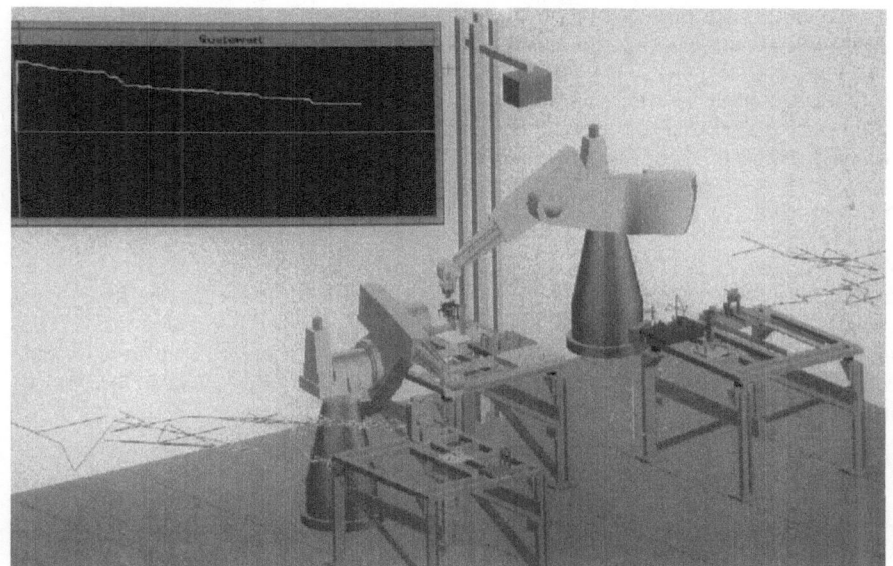

Fig. 3: Automatic Path Planning with USIS

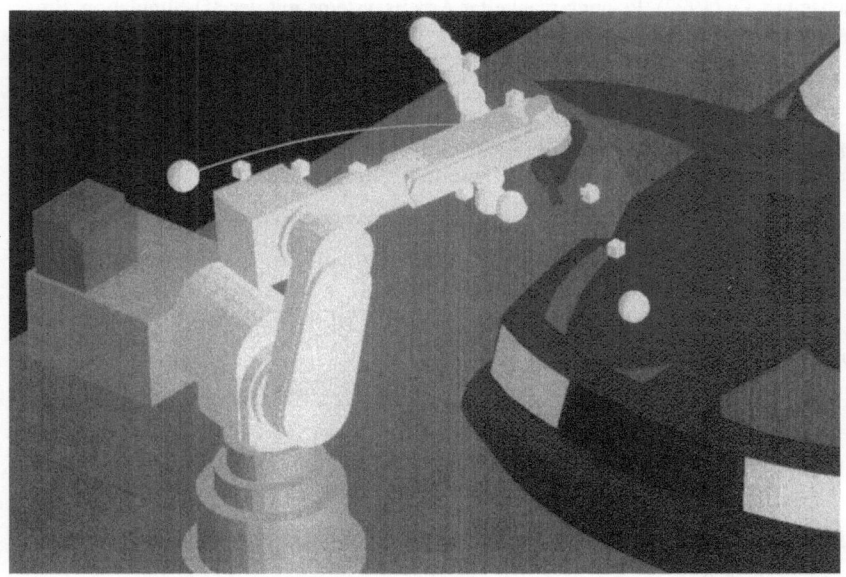

Fig. 4: Optimization Progress after 100 Iterations

configuration is estimated by the execution of the modified robot programs. The result from a special assessment function is a so called quality value, which describes the quality of the current layout configuration. This value is given back to the optimization algorithm.

With this system the planner is able to find optimized locations (fig. 4) for different cell components, robots or sensor systems. Some possible optimization criterions are for example the cycle time or the sum of angle to cover.

6 Conclusion

Advanced three-dimensional planning and simulation systems should enable the simulation of complex work cells as realistic as possible. In addition to this different planning tools should support the planner with automatic functions, like path planning algorithms or numerical layout optimization packages. USIS embodies the discussed features of advanced graphic robot simulation systems. USIS has been applied to solve various problems and has proved its functionality in several industrial projects.

Acknowlegement

The work on which this report is based is sponsored by the Deutsche Forschungsgemeinschaft (DFG) within the framework of the special research projects SFB 331 and SFB 336

7 References:

Kummetsteiner, G. (1992). Planung manueller Arbeitssysteme mit der 3D-Simulation. Produktionsautomatisierung pa 2/92 , 34-37

Milberg, J., Schuster, G. (1989). Modellierung von Sensoren für die Montagesimulation. ZwF 84 11, 650-654

Milberg, J., Woenckhaus, C. (1992). Automatic Layout Optimization Using 3D-Simulation. Proceedings of the European Simulation Symposium, Dresden, 290-294

Schrüfer, N. (1992). Erstellung eines 3D-Simulationssystems zur Reduzierung von Rüstzeiten bei der NC-Bearbeitung. Springer-Verlag, Berlin

Stetter, R. (1990). Bahn nach Plan, Roboter, mi-Verlag

Stetter, R. (1992). Einbindung physikalischer Effekte in die 3D-Simulation eines Handhabungsprozesses, Robotersysteme 8, 134 - 138

Stetter, R. (1993). Rechnergestützte Simulationswerkzeuge zur Effizienzsteigerung des Industrieroboteinsatzes, iwb-Forschungsbericht 62, 1993

Woenckhaus,C., Stetter, R., Rockland, M. (1993). Optimierung flexibler Montagezellen durch die 3D-Simulation. Tagungsbericht: Simulation und Fabrikbetrieb, Aachen, 188 - 211.

Woenckhaus,C. (1990). ADAMS in robot simulation. 6th ADAMS NewsConference, Wiesbaden, Proceedings.

Woenckhaus, C. (1992). Konzeption eines Systems zur automatischen 3D-Layoutoptimierung. Robotersysteme 8, 239-244

Emulation of Articulated Robots

G. Conte, T. Leo, S. Longhi and R. Zulli
Dipartimento di Elettronica ed Automatica
Università di Ancona

Abstract. This paper reports on the outgrowth of a complete simulation system for robotic manipulators. The development of general procedures for the kinematic and dynamic simulation of articulated robots and the integration of these procedures into a CAD system are discussed. Particular attention is devoted to the solution of the inverse kinematic and of the motion planning problem. The integration of procedures for the kinematic simulation and the preliminary studies for the integration of dynamics and control procedures confirm the importance of CAD as a basic tool for developing robotic applications.

1. Introduction

Emulation and off-line programming of robots can improve the overall efficiency of automated factories. Off-line programming by means of a CAD system (Ravani, 1988) allows robots to remain on line, performing manufacturing tasks, while being programmed for a different job. Emulation of articulated robots is also of invaluable help to test the performances of control and motion planning algorithms, in particular when there are hard constraints on the tasks to be performed.

This paper summarises the researches carried on by the authors under the Italian project "Progetto Finalizzato Robotica" of the C.N.R. in recent years. The research effort was aimed principally at the development of efficient, reliable and general procedures for the kinematic (Fioretti et al., 1990), (Conte et al., 1992) and dynamic (Fioretti et al., 1991) simulation of articulated robots and at the integration of these procedures in a CAD system.

In this context a general procedure for the automatic generation of the dynamic models of robots has been faced by means of a symbolic manipulation approach. Simplification and approximation procedures have been developed for reducing the complexity of the dynamic equations of the robots. The procedures for the automatic generation, simplification and approximation of the dynamic model have been tested making reference to a PUMA robot. The developed tools for the kinematic simulation of articulated robots have been integrated in the CAD package developed by ECOCAD, in the framework of the project mentioned above. These tools allow the definition of articulated robots with an arbitrary geometry, according to the Denavit-Hartenberg notation (Fu et al., 1987), and the derivation of the corresponding numeric inverse kinematic models. In this way it is possible to simulate the motion of articulated robots, given the curves to be followed by the end-effectors.

A general procedure for the solution of the motion planning problem for articulated robots has been developed and integrated in the CAD system.

The overall system has been successfully used for the simulation of the motion of an articulated robot in a moderately cluttered environment. A further application deals with two robots acting in a common workspace. The described procedures are written in C language, in X/Windows environment. Software portability has been effectively achieved in this way.

The paper is organised as follows. Section 2 describes the algorithms used for kinematic simulation. Section 3 describes the dynamic modelling of the robotic manipulators. In section 4 it is described the proposed solution to the motion planning problem. Section 5 reports on considerations about the overall integrated simulation system, describing the planned enhancement of the system.

2. Kinematic model

In order to solve in a general manner the kinematic problem, the generation of the kinematic model has been done in symbolic (Fioretti et. al., 1990) and numeric form. The class of robots that has been considered is characterised by open kinematic chains of rigid bodies: only non-redundant chains are considered.

In order to solve the inverse kinematic problem the differential relationship between the vector r, that describes the position and orientation of the end-effector and the vector q of the joint variables has been considered. This relation is described by the following linear equation with the Jacobean matrix $J(q)$ as the coefficient matrix:

$$dr = J(q)dq. \qquad (2.1)$$

In the present case, $J(q)$ is a $n \times n$ matrix almost everywhere non-singular hence the solution of the inverse kinematic problem is straightforwardly obtained by inversion of $J(q)$. In the singular points, where $J(q)$ is singular, and in the respective neighbourhoods, the preceding solution does not work. Consequently, proper pseudo inverses (Nakamura and Hanafusa, 1986) have to be considered. In this way it is possible to obtain continuous solutions that are feasible even in correspondence and in the neighbourhoods of the singular points.

The automatic generation of the direct and inverse kinematic models has been performed utilising a symbolic manipulation package (REDUCE) so that the inverse and the pseudo inverse of the $J(q)$ matrix can be explicated in symbolic form. The input data for the above symbolic procedures are provided giving a geometric description of the kinematic chain. These data refer to the Denavit-Hartenberg notation and are relative to the type of joints (prismatic or revolute) and to the link dimensions. They are directly obtainable by the geometric model of the manipulator, as represented in the CAD system. The curve to be followed by the end-effector of the manipulator is defined by means of the CAD system and, using the developed procedure, it is possible to simulate the motion of the manipulator. The system tests automatically the feasibility of the given trajectory.

3. Dynamic model

The dynamic model of an n-degrees of freedom manipulator is described by:

$$M(t) = B(q(t))\ddot{q}(t) + C(q(t))h(\dot{q}(t)) - E(q(t)) \qquad (3.1)$$

where $M(\cdot) \in R^n$ is the vector of generalised forces, $B(\cdot)$ is the $(n \times n)$ inertia matrix, $C(\cdot)$ is the $(n \times (n+1)/2)$ matrix of the centrifugal and Coriolis terms, $E(\cdot) \in R^n$ is the vector of the potential terms and $h(\cdot) \in R^{n \times (n+1)/2}$ is the vector of all the quadratic terms of generalised velocities. The automatic generation of the dynamic model, in symbolic form, has been performed using the DYMIR program (Vecchio et. al., 1980). Using the symbolic approach, the numerical evaluation of the model equations for simulation and control applications is highly time-consuming. The existence of this drawback is principally due to the lack of simplification during the symbolic calculations. It is possible to reduce the complexity of the dynamic model by means of simplification and approximation procedures (Fioretti et al., 1991). By simplification of model expressions equivalent but simpler expressions are obtained. The approximation of model expressions leads to non-equivalent expressions, having omitted negligible terms. Both approaches have been pursued. The developed packages contain procedures for the translation of the obtained symbolic expressions in statements acceptable by numerical programming languages.

These procedures help in testing the dynamic characteristics of a manipulator when there are strict constraints both on the accuracy of the positioning and on the rapidity of the movement. The developed tools have been tested using as reference model a PUMA 560 manipulator.

4. Motion planning

In order to have a complete simulation system it is necessary to develop procedures for the automatic generation of collision-free paths. This problem, known in its simplest form as the motion planning problem, is also a major problem that one needs to solve on the way of creating autonomous robots. Theoretical considerations (Schwartz et al., 1987; Canny, 1987) allow to state that the complexity of algorithms for the solution of the motion planning problem is at least exponential in the number of degrees of freedom (DOF). Multiple robotic systems, such as the co-operative systems that it is proposed to simulate, are characterised by an high number of DOF and thus the difficulty in the solution of the motion planning problem is greatly enhanced.

A general procedure for approaching the motion planning problem can be based on the Cspace setting (Lozano-Perez, 1981, 1983) and on a collision check procedure like that described in (Hwang, 1990). The Cspace, in which the motion of the manipulators is represented by that of a single point, is constructed using an approximate cell decomposition method, as described in (Lozano-Perez, 1983), and, by means of the so-called Wavefront Expansion Algorithm (Latombe, 1991), it is endowed with a distance field (Borgefors, 1986), (Conte et al., 1992). Then a path for the point representing the manipulators in the Cspace can simply be found following the negative gradient of the distance field. The found path is the shortest feasible path between the given start and goal positions.

The main problem with the above method is that the Cspace complexity grows exponentially with the number of DOF of the problem (Gupta, 1992). Using a centralised planning approach and representing the manipulators as a single point in a global Cspace, the typical number of DOF is included between six and ten. Current literature (Latombe, 1991) reports the practical possibility of using an approximate cell decomposition method only for a number of DOF less or equal to four.

A general strategy has been developed for reducing the complexity of the problem and for limiting the amount of memory needed to represent the Cspace. In this way it is possible to solve problems for robots with many degrees of freedom. The effectiveness of the multilevel discretization in the reduction of Cspace complexity has been tested in the case of two 3-DOF manipulators acting in the same environment.

5. Concluding remarks

The research carried on by the authors in the framework of the Italian project "Progetto Finalizzato Robotica" of C.N.R. was aimed at the development of a suitable software tool for the kinematic and dynamic simulation of robotic manipulators. This software tool was meant to be used for testing the performances of control algorithms and the effectiveness of path and trajectory planning algorithms.

A kinematic simulation tool has been developed and integrated in the CAD package ECOCAD. This software tool allows the definition of a robot with an arbitrary geometry, according to the Denavit-Hartenberg representation, and the derivation of the corresponding inverse kinematic model. A number of widely used manipulators, such as the Stantford and the Elbow ones, have been constructed by means of the CAD system and successively used for simulation. Therefore, it is possible to visualise the motion of the manipulators, given the curves to be followed by the end effectors.

Furthermore a collision-free path planning algorithm has been developed and integrated in the same CAD system. Obstacles of arbitrary geometry and dimensions can be defined in the workspace of the manipulators: the corresponding set of forbidden configurations is automatically computed. The approach used to solve the motion planning problem is a complete one because our procedure always finds a collision-free path, whenever such a path exists. The described tool for path planning has been applied to plan the motion of two 3-DOF manipulators acting in the same environment.

Further developments of the proposed simulation system include the integration of dynamic and control algorithms in the CAD system and the simulation of a complex existing workcell consisting of two J0 MACH 16B manipulators designed to work on huge elements.

6. Acknowledgements

This work was supported by Research Grants of C.N.R. in the context of Progetto Finalizzato Robotica, contracts no. 90.00365.PF67, 91.01930.PF67 and 92.01072.PF67.

7. References

Borgefors, G. (1986). Distance transform in arbitrary dimensions. In: Computer vision, graphic image processing, vol.34, pp. 344-371.

Canny, J.F. (1987). The complexity of robot motion planning. MIT Press.

Conte, G., Longhi, S., Zulli, R. (1992). An algorithm for CAD-based generation of collision-free paths for robotic manipulators. In: Preprints of IFAC Workshop CIM'92, Wien, Austria.

Fioretti, S., Jetto, L., Longhi, S. (1990). Geometric modelling of non-redundant manipulators: integration with kinematic and dynamic modelling. In: Mc Graw Hill Book (Ed.) Proceedings of ICARCV'90, Singapore, pp.616-625.

Fioretti, S., Leo, T., Longhi, S., Pepe, P. (1991). Complexity reduction of robot dynamic equations. In: Proceedings of IEEE TENCON'91, New Delhi, pp.307-311.

Fu, K.S., Gonzales, R.C., Lee, C.S.G (1987). Robotics: control, sensing, vision and intelligence. Mc Graw Hill, Singapore.

Gupta, K.K. , (1992). Motion planning for many degrees of freedom: sequential search with backtracking. In: Proceedings of IEEE International Conference on Robotics and Automation, pp.2328-2333.

Hwang, Y.K. (1990). Boundary equations of configuration obstacles for manipulators. In: Proceedings of IEEE International Conference on Robotics and Automation, pp. 298-303.

Latombe, J. (1991). Robot motion planning. Kluwer Academic Press.

Lozano-Perez, T. (1983). Spatial planning: a configuration space approach. In: IEEE Transactions on Computers, Vol.C-32, pp. 108-120.

Lozano-Perez, T. (1981). Automatic planning of manipulator transfer movements. In: IEEE Transactions on System, Man, and Cybernetics, Vol.SMC-11, pp. 681-698.

Nakamura, Y., Hanafusa, H. (1986). Inverse kinematic solutions with singularity robustness for robot manipulator control. In: Journal of Dynamic systems, Measurement and Control, vol.108, pp.163-171.

Ravani, B. (1988). CAD based programming for sensory robots. NATO ASI series F50, Springer Verlag, Berlin.

Schwartz, J.T., Sharir, M., Hopcroft, J. (1987). Planning, geometry and complexity of robot motion. ABLEX publishing.

Vecchio, L., Nicosia, S., Nicolò, F., Lentini, D. (1980). Automatic generation of dynamic models of manipulators. In: 10th International Symposium on Industrial Robots, Milan.

ALMORO - Symbolic Robot Dynamics Customizer

K. Swider

Technical University of Vienna, Institute for Handling Devices and Robotics, Vienna, Austria and
Technical University of Rzeszow, Department of Electrical Engineering, Rzeszow, Poland

Abstract. The objective of the paper is to demonstrate some application of computer algebra methods and tools in robotics through the development and implementation of symbolic manipulation package ALMORO for robot dynamics modelling and customizing. The tremendous size of explicit dynamic model is the great barrier first to its derivation and second to its customization for use in model-based control algorithms. The aim of the paper is to show that the considerable reduction of computational requirements could be achieved through systematic numerical simplifications. The method proposed bases on the iterative search procedure using symbolically generated set of simplification strategies. The simplification errors are numerically verified to be comparable with common modelling uncertainties.

1. Introduction

The present day industrial robots are usually very complicated mechanical systems whose dynamics is described by the highly coupled and non-linear, second-order differential equations. The base equation for the *closed-form* dynamic model of an open chain robot with N rigid links (DOF) is

$$\mathbf{D}(\mathbf{q})\ddot{\mathbf{q}} + \mathbf{C}(\mathbf{q},\dot{\mathbf{q}}) + \mathbf{G}(\mathbf{q}) = \mathbf{F}(t) \tag{1}$$

It is the system of N differential equations with following coefficients: $\mathbf{D}(\mathbf{q})$- inertial coefficient matrix (NxN), $\mathbf{C}(\mathbf{q},\dot{\mathbf{q}})$- centrifugal and Coriolis forces vector (N) and $\mathbf{G}(\mathbf{q})$- gravitational forces vector (N) which are functions of the instantaneous configuration in state space of generalized joint coordinates $\mathbf{q}$ and velocities $\dot{\mathbf{q}}$. The detailed specification of these parameters and their physical and mathematical properties have been widely presented in robotic literature (Paul,1981), (Tourassis and Neuman, 1985). In (1) $\mathbf{F}(t)$ is a vector of actuating forces/torque's (N). It requires a fair amount of human efforts to obtain the analytical formulas for $\mathbf{D}$, $\mathbf{C}$ and $\mathbf{G}$ and therefore *symbolic computations* (computer algebra) are often applied for automatic generation of robot dynamic models (Murray and Neuman, 1984), (Burdick, 1986), (Swider,1992,2). However the on-line evaluation of robot dynamics is computationally expensive and time consuming task and therefore requires the use of numerically simplified models in controllers. The sophisticated attempts to simplify the robot model are usually based upon analysis of numerical significance of separate dynamic coefficients rather then model properties and fundamental principles of classical mechanics. The need of systematic simplification procedures has been postulated in literature (Tang and Tourassis, 1989).

2. Numerical simplifications of dynamic models

2.1 Numerical analysis of simplified models

In order to perform numerical simplifications the automatically generated *symbolic* closed-form model is considered (Murray and Neuman, 1984). After replacing the constant parameters by numerical values we obtain *symbolic-numeric* form. The goal of simplification procedure is to determine which of terms of the coefficients in symbolic-numeric form can be neglected in simplified model. The following method is based on numerical analysis of simplified models. The simplification errors are figured for all model coefficients in the joint coordinates space and for typical trajectories as well. The analogous errors are estimated for inertia matrix determinants for actual and simplified models giving the possibility to prevent the inherent property of the dynamic robot model as positive-definiteness of inertia matrix (Tourassis and Neuman, 1985). As the result of error analysis the following quantities are obtained (Swider,1992,2):

- the simplification errors for each model coefficient and joint force/torque
- the percentage reducing of computational requirements for dynamics evaluation.

Additionally the numerical analysis provides the possibility to estimate the simplification tolerance for each coefficient. The idea of this estimation is to assume the simplification errors to be comparable with common modelling uncertainties (measurement tolerances of model parameters, unmodelled friction etc).

2.2 Simplification procedure

The simplification procedure is accomplished in two phases. On the first stage the simplification errors are estimated using numerical analysis as mentioned above. The estimated values are then used on stage two where the numerical simplifications are carried out. This stage starts with assigning the common admissible domain for model simplifications that follow. Given arbitrally some modelling tolerance, which could be represented as a number ε ($0<\varepsilon<1$), a set of possible simplification strategies is generated (Swider,1992,1). Than the iterative searching procedure is applied to find the strategy which satisfies the following two requirements:

- the simplification errors do not override the estimated values
- the positive-definiteness and symmetry of inertia matrix are preserved in simplified model.

In comparison to common methods known from literature (Tang and Tourassis, 1989) the improvements of the method proposed are:

- for all model coefficients the modelling tolerance is found individually
- the main inherent properties of dynamic models can be preserved
- the agreeable threshold of numerical simplification can be estimated.

A LISP language is very well adapted to symbolic manipulation and has been used commonly in symbolic model generation (Murray and Neuman, 1984), (Burdick, 1985). In algebraic robot modeler - ALMORO - developed by the author the symbolic computation in LISP have been applied to automatically simplify the symbolic-numeric models and to generate the procedures for numerical analysis of simplified models.

3. An example

The dynamic model of 3 DOF (positioning system) of PUMA robot has been simplified to illustrate the performance of the iterative systematic simplification procedure. On the first stage the admissible simplification errors have been estimated for each coefficient assuming the 5% tolerance in each of the inertia parameter values. With this assumption the estimated error e.g. for the inertia (mass) matrix coefficient d_{11} was 7.14% w.r.t. its range. The set of simplification strategies has been generated for $\varepsilon=0.4$ and the satisfying simplification strategy was found after 3 steps of the iterative procedure as shown in Tab.1. On each step

Tab.1. PUMA 560 model simplification

ITERATION	MODEL REDUCTION (%)	RELATIVE ERROR OF d_{11}
I	(70.93 63.33 60.29)	26.27
II	(47.67 50.67 42.65)	17.14
III	(38.37 39.33 33.82)	2.11

the actual and simplified models have been numerically analyzed and the errors occurred were compared with corresponding admissible errors (for d_{11} these errors are indicated in the third column of Tab.1).

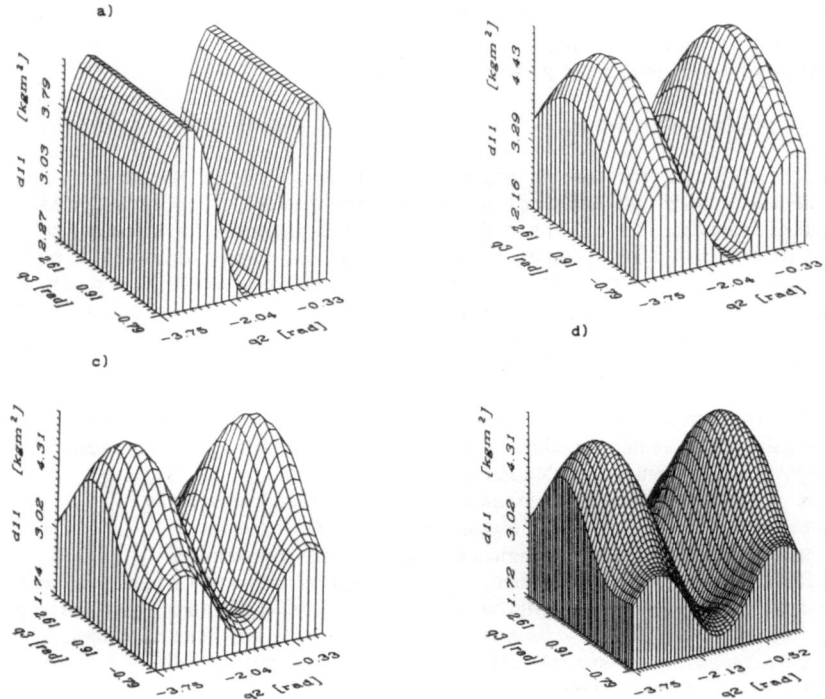

Fig.1. Simplifications of d_{11} ; a) iteration I b) iteration II c) iteration III d) actual

The percentage reduction of operations in model evaluation (additions, multiplications, trigonometric functions) are also computed and listed in column 2. In the case of 3 DOF the coefficient d_{11} can be plotted as function of two joint coordinates q_2 and q_3 (Fig.1). While applying the consecutive steps of simplification procedure the shape of the simplified coefficient becomes successively closer to the actual.

4. Conclusions

A number of computer programs have been implemented and presented in literature to derive explicit robot dynamics models. ALMORO - implemented by the author - symbolically generates the forward solution and complete robot dynamics model using Lagrangian formulation. In this paper symbolic computations have been used to *obtain the numerically simplified models* by applying the systematic simplification procedure. The special options have been implemented in ALMORO to perform automatically the particularly complex and time consuming operations as generating the set of simplification strategies, model simplifications according to given strategy etc. The automatic generation of the procedures for numerical evaluation enables to compare the performance of actual and simplified models for existing manipulators. The results are very encouraging (Swider, 1992,2).

5. References

Burdick, J. (1986). An algorithm for generation of efficient manipulator dynamic equations. In: Suri, R. (Ed.). Proc. IEEE Int. Conf. Robotics and Automation, San Francisco, pp. 212-218

Murray, J.J., Neuman C.P. (1984). ARM: An algebraic robot dynamic modelling program. In: Paul , R.P. (Ed.). Proc. 1-st Int. IEEE Conf. Robotics, Atlanta, pp. 103-114

Paul, R.P. (1981). Robot Manipulators: Mathematics, Programming and Control. MIT, Cambridge

Swider, K. (1992). An automatic verification of simplified robot models. In: Tzafestas, S.G. (Ed.). Robotic Systems. Advanced Techniques and Applications. Kluwer Academic Publishers, pp. 53-60

Swider, K. (1992). Automatic generation and reduction of symbolic models of robot dynamics. Ph.D. dissertation. Department of Electronic Engineering, Warsaw Institute of Technology, Warsaw

Tang, K.M.W., Tourassis, V.D. (1989). Mathematical deficiency of numerically simplified dynamic robot models. IEEE Trans. on Autom. Control, 10

Tourassis, V.D., Neuman C.P. (1985). Properties and structure of dynamic robot models for control engineering applications. Mech. Machine Theory, 20, pp. 27-40

Techniques to Improve the Performances of an Industrial SCARA Robot

R. Faglia, C. Remino
Mechanical Engineering Department
University, Brescia, Italy

Abstract: This paper suggests some techniques to improve the performances of industrial SCARA robots. Such techniques are: 1) a method of calibration based on optical measurement, which increases the robot accuracy; 2) a parametric definition of the actuators motions, which optimizes the dynamic behaviour of the robot.
Both techniques can be applied on-field, when the manipulator is operative.

1. A calibration procedure for increasing the robot accuracy

The precision of a robot depends on the structural errors, that is the errors that affect the parameters of Denavit and Hartemberg, such as the length of the links and the alignment of the joint axes. Such errors are constant and their effect on the position of the gripper is repetitive. A possible way to improve the accuracy of a robot is to compensate for the structural errors with software procedures. This approach is far more cost-effective than manufacturing a "perfect" robot and can improve the accuracy of manipulators that already work.
The researches carried out by the authors (Faglia and others, 1992a; Faglia and others, 1993) can be summed up in the following four steps:

1. to achieve relationships that express the pose errors of the gripper in function of the structural errors;
2. to build an algorithm for compensating for the errors (see also Cai and Rovetta, 1988);
3. to set up a method for measuring the pose of the gripper with optical instruments (see also Nowrouzi and others, 1988);
4. to use these procedures to calibrate an industrial SCARA robot (IBM 7535).

Be: S the set of gripper coordinates (usually 6) in the robot workspace; Q the vector of the n joint coordinates; L the set of the m structural parameters (the structural parameters of a robot with R rotational degrees of freedom and P prismatic d.o.f. are m = 6 + 4 R + 2 P (Mooring and others, 1991)). The position of the gripper S depends on the structural parameters L and the joint coordinates Q by the direct kinematic relation:

$$S = F(Q, L) \qquad (1)$$

Be L_n the set of nominal values of the structural parameters and suppose the structural errors are small. In this hypothesis, the gripper reaches the "real" position S_r:

$$S_r = S + \Delta S \cong F(Q, L_n) + \frac{\partial F}{\partial L} \Delta L \qquad (2)$$

where ΔS is the positioning error due to the structural error ΔL. ΔS and ΔL depend each other by the following equation:

$$\Delta S \cong J_L \Delta L \quad \text{with} \quad J_L = \frac{\partial F}{\partial L} \qquad (3)$$

where J_L is the Jacobian matrix of the system (1). J_L depends on Q and L, and must be evaluated considering the robot free from both structural and position errors (L_n and Q respectively).

Equation (3) states that, for a given configuration **Q** of the joint variables, the pose error is "proportional" to the structural error ΔL. ΔL can be found moving the gripper to some known positions $(Q_1, Q_2, ..., Q_j)$ and measuring the pose errors that occur in one or more coordinates $(\Delta S_1, \Delta S_2, ..., \Delta S_i)$. Since m unknowns must be identified, at least m different measurements have to be carried out (Faglia and others, 1992a; Faglia and others, 1993).

If the theoretical coordinates **Q** get little changes ΔQ, the gripper will reach the new position S_c:

$$S_c \cong F(Q, L_n) + J_L \, \Delta L + J \, \Delta Q \quad \text{with} \quad J = \frac{\partial F}{\partial Q} \qquad (4)$$

where **J** is the "usual" jacobian matrix of the robot. Thanks to Equation (4) we can compensate for the structural errors modifing the value of the joint coordinates by the following amount:

$$J_L \, \Delta L + J \, \Delta Q \cong 0 \quad \Rightarrow \quad \Delta Q \cong -J^{-1} J_L \, \Delta L \qquad (5)$$

This method has been used to calibrate an IBM 7535 SCARA robot. During the tests the orientation and the position of the gripper have been measured by a laser triangulation (Faglia and others, 1992a) (Fig.1) and a commercial interferometer (HP5527A) (Fig. 2) respectively.

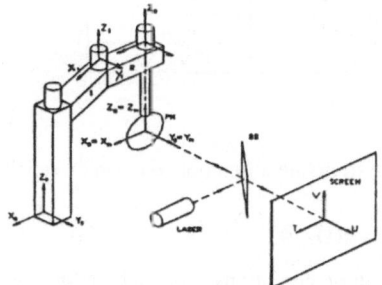
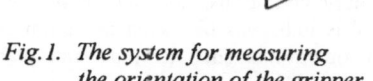

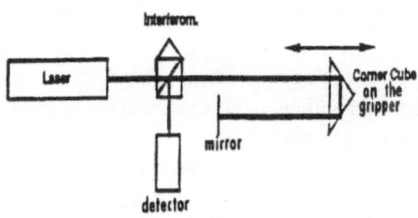

Fig.1. The system for measuring the orientation of the gripper.

Fig. 2. The system for measuring of the gripper.

The position errors have been compensated for by a software program that:

1. reads the characteristics and the previously evaluated structural errors of the robot from a data base;
2. receives, as input, the task that a "perfect robot" (without structural errors) should perform in term of gripper displacement (first column of Table 1);
3. provides, as output, the new path (fourth column of Table 1) to be carried out by the robot to reach the right position.

2. Trajectory planning for optimal movement

The motion (or trajectory) planning is the search of a strategy for moving an actuator according to a pre-fixed task. In this regard the authors have aimed not only to build some theoretical motion planning algorithms, but also to apply them to actual cases.

The shown method is based on a parametric definition of the actuators motion profile: changing the parameters which define the movement, it is possible both to fulfil the pre-fixed task and to optimize a pre-defined functional.

Suppose that we want to describe the movement of an actuator that displaces of Δx in a time **T**,

	Theoretical position	Before calibration	After calibration	Path for compensation
x	100.00	92.25	100.41	107.74
y	300.00	307.18	300.11	292.82
z	300.00	295.56	300.01	304.43
x	100.00	108.98	100.39	91.02
y	350.00	343.25	350.12	356.74
z	300.00	304.67	300.01	295.32
x	100.00	110.28	100.37	89.71
y	400.00	393.78	400.13	406.21
z	300.00	304.89	300.00	295.11
x	100.00	111.64	100.35	88.36
y	450.00	393.78	450.14	455.58
z	300.00	304.89	300.00	294.93
x	100.00	113.06	100.31	86.94
y	500.00	495.19	500.15	504.40
z	300.00	305.21	300.00	294.79
x	300.00	295.80	300.36	295.80
y	300.00	292.41	299.98	307.58
z	300.00	304.45	300.04	395.55

Table 1

starting and ending its movement with null velocity. We have to define a function **x(t)** which satisfies the following constraints:

$$x(0) = 0 \qquad x(T) = \Delta x \qquad \dot{x}(0) = \dot{x}(t) = 0 \qquad (6)$$

Note that the movement has been defined only by boundary conditions, so such definition is inadeguate for applications where a continuous path control is indispensable, such as continuous welding operations and painting, while it is suitable for operations such as pick and place, assembling, etc.

One of the possible ways to define a displacement is to sum up a function that meets the given boundary conditions with harmonic functions having frequencies multiple of 1/T and null value at time **0** and **T** (Schmitt and others, 1985):

$$x(t) = \frac{1}{2} \Delta x \left(1 - \cos \pi \frac{t}{T} \right) + \sum_{k=1}^{n} \frac{1}{2} \Delta x \, p_k \left[1 - \cos \left(2k\pi \frac{t}{T} \right) \right] \qquad (7)$$

where: **n** is the number of the harmonic functions; p_k is the amplitude (or weight) of the k-th harmonic; **t** is the time variable. It can be easily proved the boundary conditions (6) are fulfilled for any combination of the weights p_k, which affect only the shape of the law of motion.

Now, if a functional **F** depending on the motion shape (that is **x(t)** and its derivatives) is settled, it will be possible to minimize (or maximize) **F** changing the weights p_k. This search for the right weights can be done with an optimisation algorithm for multiple variables (for example the Pattern Search one (Press and others, 1986)).

This method has been used to increase the dynamic behaviour of an industrial SCARA robot, an ICOMATIC SCARA 03. As the acceleration peaks are important values to judge both the motion and the inertial behaviour of the system, this quantity has been assumed as the functional to optimize. The task imposed to the robot was to rotate the first link of 45 degrees in 0.77 sec, keeping the forearm aligned with. The gripper was uncharged.

By the optimization procedure the law of motion in Fig. 3 has been found: it is the one which minimizes the negative acceleration peak of the gripper during the movement.

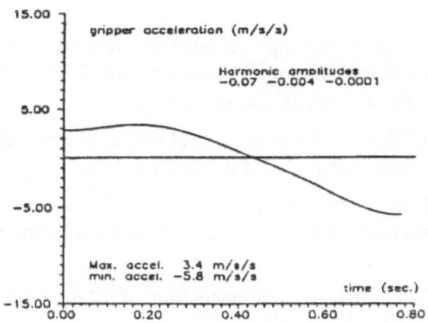

Fig.3 The motion profile which minimizes the negative acceleration peak of the gripper.

Figures 4 and 5 show the acceleration of the gripper when the bang-bang profile and then the optimized one has been imposed to the motors: the change in the profile has decreased the minimum acceleration peak, which passes from 10.4 m/sec^2 in the bang-bang law of motion (Fig. 4), to 7.2 m/sec^2 in the optimized one (Fig. 5).

This test has been performed on other movements and with different functionals, always giving satisfactory resu ts (Faglia, 1992c).

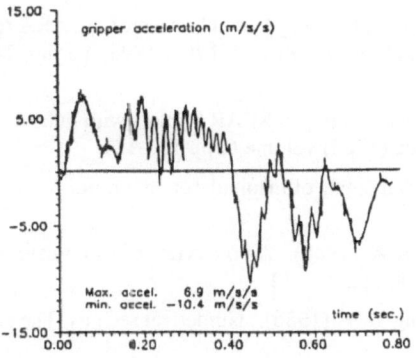

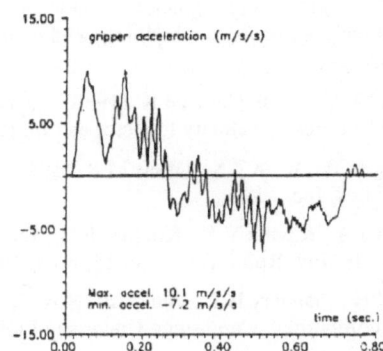

Fig.4 The acceleration of the gripper in the bang-bang law of motion.

Fig.5 The acceleration of the gripper in the optimized law of motion.

3. Conclusions

The paper has shown some experimental results obtained with a calibration method based on optical measurements and a movement optimisation procedure. The main characteristic of such procedures is that they can be easily used on-field, when the robot is already functioning. In other words, they allow to improve the performances of the robot without changing its architecture.

The procedures have been tested on SCARA robots, but can be generalised for and extended to other servo-systems.

4. Bibliography

- Cai M.S., Rovetta A. (1988). Linear and angular position, velocity and acceleration error analysis for robot manipulators and their compensation, Proccedings of IEEE Intern. Workshop on Intelligent Robot and Systems '88, Tokyo, Japan, Oct.1988.

- Denavit J., Hartenberg R.S. (1955). A kinematic notation for lower-pair mechanisms based on matrices, In: Trans. ASME J. Appl. Mech., 22, pp.215-221, June.

- Faglia R., Legnani G. (1989). Dynamic Analysis of SCARA Robots for Optimal Motions with Low-Cost Controller. Proc. Second Congress IFAC "Low-cost Automation" 8-10 November 1989, Milan, Italy.

- Faglia R., Legnani G. (1991). On the dynamic behaviour of robots with non-rigid transmission. Proc. VIII IFToMM, Praha, 26-31 August 1991.

- Faglia R., Legnani G., Docchio F., Minoni U. (1992a). Experimental evaluation of the joint axes misalignment in a SCARA robot by optical measurements. In: Mechatronic Systems Engineering, Kluwer Academic Publishers, vol.1, 1992, pp.301-314.

- Faglia R., Legnani G. (1992b). Harmonic Drive transmissions: the effects of their elasticity, clearance and irregularity on the dynamic behaviour of an actual SCARA robot. In: Robotica (1992) Volume 10, pp 369-375.

- Faglia R. (1992c). Trajectory shaping for flexible manipulators: an optimization algorithm and an application to a SCARA robot. Proc. 3rd International Workshop on Advances in Robot Kinematics 3rd ARK, September 7-9, 1992, Ferrara, Italy.

- Faglia R., Legnani G., Magnani P.L., Docchio F., Minoni U. (1993). Calibration of SCARA robot by optical measurements: methodology and experimental results. Proc. ISMCR 1993, Torino, 21-24 Sept. 1993.

- Ibrahim M. Y., Cook C., Tieu K. (1988). Dynamic behaviour of a SCARA robot with links subjected to different velocity trajectories. In: Robotica (1988) volume 6, pp.115-121.

- Mooring B.W., Roth Z.S., Driels M.R. (1991). Fundamentals of manipulator calibration. John Wiley & Sons, Inc., USA.

- Nowrouzi A., Kavina Y.V., Kochekali H., Whitaker R.A. (1988). An overview of robot calibration techniqes. In: Ind. Robot (UK), vol.15, no.4, Dec.1988, pp.229-232.

- Press W.H., Flannery B.P., Teuklosky S.A., Vetterling W.T. (1986). Numerical recipes. The art of scientific computing. Cambridge University Press.

- Schmitt, Soni, Srinivasan, Naganathan (1985). Optimal Motion Programming of Robot Manipulators. In: ASME Trans. Journal of Mechanisms, Transmissions and Automation in Design, Vol.107/239, June 1985.

5. Acknowledgments

The MIVAL-ICOMATIC S.p.A. Gussago Brescia and the IBM SEMEA, Vimercate (Milano) are acknowledged.

Work supported by CNR 9201076 Pos. 11610746 grants

Experimental Multiprocessor Robot Controller

M. Terbuc, K. Jezernik
Laboratory of Industrial Robotics
Faculty of Technical Sciences - ERI, University of Maribor,Slovenia

Abstract: It is very important, for pedagogic and research purposes, that a robot controller is easy to use and its software is simple to modify. Our experimental multiprocessor laboratory robot controller is an open system, and with the transputers used as central processors a great processing power and a very fast communication were achieved. In the Robotics Laboratory at the Faculty of Technical Sciences in Maribor we developed new control concepts which assure trajectory tracking and the incorporation of the external sensors. Standard controllers compute inverse transformation first and then execute the control in joint coordinates for a single robot axis. We used an advanced control scheme in Cartesian space with included kinematics and dynamics. In this way we obtained a better tracking robustness. The multiprocessor structure allows computing time of 1ms.

1. Tracking Control in the Cartesian Space

Conventional robot control techniques are based on a separate treatment of the kinematic and dynamic structure. Inverse kinematics transforms task space into joint coordinates where dynamic motion is executed. This robot system control is improper because naturally connected processes are treated separatly. Tracking tasks are specified in task space and also trajectory planning. Today we also expect from robots to achieve a high tracking accuracy at a momentary payload change. The structure of robots is very nonlinear because of coupling between axes, so we use robust tracking control, which is independent of parameter variations and disturbances. A natural dynamic control scheme is in the task space, which has a lot of advantages in comparison to used joint coordinates schemes.

Most of the robot aplications involve a task specified in the task space (Cartesian). The basic issue in a sensor-based robot control problem can be categorized as: Task specification, Trajectory planning, Sensory data acqusition and interpretation and Close loop control. Every loop of the control algorithm must incorporate each of the above steps. The closed loop control can be purely kinematic or based on a dynamic model. Here we are concerned with the dynamic model.

The nonlinear feedback based task space servo controller is a controller of this kind. The development of such a controller is discussed in (Jezernik et. al., 1992).Let

$$D(q)\ddot{q} + h(q,\dot{q}) = \tau$$
$$y = L(q) \tag{1}$$

denote a nonredundant robot model together with the output. Here $D(q)$ is a(6x6) Inertia matrix, $h(q,\dot{q})$ is a (6x1) vector of the Coriolis, Centrifugal and Gravity terms, τ a (6x1) vector of the joint torques and y a (6x1) vector of output. The nonlinear feedback based task servo controller for the trajectory tracking closes the feedback loop according to:

$$\tau = D(q)J_h^{-1}(q)[v - \dot{J}_h\dot{q}] + h(q,\dot{q}) \tag{2}$$

where $v = \ddot{y}^d + K_v(\dot{y}^d - \dot{y}) + K_p(y^d - y)$. Here q is a (6x1) vector of the joint positions. K_v and K_p are diagonal (6x6) velocity and position feedback matrices. The superscript ^{d} denotes the desired trajectory variables. If we define $L(q) = q$ then the controller that tracks the same end-effector path as above, is:

$$\tau = D(q)[\ddot{q}^d + K_v'(\dot{q}^d - \dot{q}) + K_p'(q^d - q)] + h(q,\dot{q}) \tag{3}$$

Here q^d, $\dot{q}^d$ and $\ddot{q}^d$ are derived from the Inverse kinematics and Inverse Jacobian (i.e. J_h^{-1}). The servo torque is in the joint space. Both of the model based servo schemes are essentially exact linearization and decoupling schemes with servo design in a linear system framework. If a completely preplanned end-effector trajectory is available, then the inverse kinematics and the inverse output Jacobian can form the output of a joint level planner and hence a joint servo scheme becomes computationally less expensive than a task servo scheme. The task servo scheme has many qualitative benefits:

- It is more general in its theoretical development.
- It is more intuitive in the nature and the task planner interacts more directly with the controller.
- As robot tasks are specified in the task space, it conforms to the true meaning of the word servo.
- Differential Geometry provides a natural tool for the characterization of curves in space and hence fits conceptually into the scheme. Thus, despite a computational burden, the task space servo scheme has several advantages.
- It is a step towards "Intelligent Control". It is not always possible to get completely preplanned trajectories. Moreover, intelligent planners are likely to be event-based, rather than explicitly time-based. Additional torque constraints may affect trajectory planning. In such a scenario both schemes entail the same computational steps, but different levels in the computational flow. The difference would be inverse kinematics in the joint servo schemes and forward kinematics in the task servo scheme.

Tracking control in the Cartesian space was used in order to improve the trajectory tracking and to incorporate the force sensors (Komada and Onishi, 1989) (figure 1). Position, velocity and acceleration in the Cartesian space represent the input data, and the reference motor torque represents the output data. Trajectory tracking is substantially improved by the inclusion of data about the desired velocity and acceleration (Pritschow et. al.,1992).

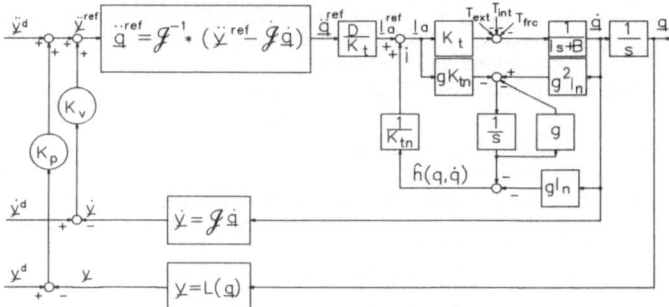

Figure1: Cartesian Space Controller

It is evident from the scheme that the direct transformation of position and velocity and the inverse transformation of acceleration must be calculated. The following equations give the direct transformation of position (Eq. 4,5), velocity (Eq. 6,7) and the inverse kinematic transformation of acceleration (Eq. 8-11) for the used Scara robot:

$$x = a_1 \cos(q_1) + a_2 \cos(q_1 + q_2) \tag{4}$$
$$y = a_1 \sin(q_1) + a_2 \sin(q_1 + q_2) \tag{5}$$
$$v_x = \dot{x} = -y\dot{q}_1 - a_2 \sin(q_1 + q_2)\dot{q}_2 \tag{6}$$
$$v_y = \dot{y} = x\dot{q}_1 - a_2 \cos(q_1 + q_2)\dot{q}_2 \tag{7}$$
$$k_1 = \ddot{x}^d + K_v(\dot{x}^d - \dot{x}) + K_p(x^d - x) + \ddot{y}\dot{q}_1 + \cos(q_1 + q_2)(q_1 + q_2)(\dot{q}_1 + \dot{q}_2\dot{g}_2) \tag{8}$$
$$k_2 = \ddot{y}^d + K_v(\dot{y}^d - \dot{y}) + K_p(y^d - y) - \dot{x}\dot{q}_1 + \sin(q_1 + q_2)(q_1 + q_2)(\dot{q}_1 + \dot{q}_2\dot{g}_2) \tag{9}$$
$$\ddot{q}_1^{ref} = \frac{1}{Det\underline{J}(q)}(-a_2 \cos q_1 + q_2 k_1 - a_2 \sin q_1 + q_2 k_2) \tag{10}$$
$$\ddot{q}_2^{ref} = \frac{1}{Det\underline{J}(q)}(xk_1 + yk_2) \tag{11}$$

The constant K_v represents velocity gains and K_p represents position errors gains.

The robot mechanism in joint space presented by (Eq. 1) with the control law (Eq. 2), precisely tracks the given trajectory and eliminates the initial conditions and disturbances due to indefinitenesses and external influences by the required dynamics of second order, given by the error dynamics:

- in joint space:
$$(\ddot{q}^d - \ddot{q}) + K'_v(\dot{q}^d - \dot{q}) + K'_p(q^d - q) = p(t) \tag{12}$$
- and in task space:
$$(\ddot{y}^d - \ddot{y}) + K_v(\dot{y}^d - \dot{y}) + K_p(y^d - y) = \mathcal{J}_h p(t) \tag{13}$$
$$(\ddot{y}^d - \ddot{y}) \doteq \mathcal{J}_h(\ddot{q}^d - \ddot{q}) = \mathcal{J}_h p(t) \tag{14}$$

Error (Eq. 14) occur only in the acceleration signal, because the acceleration signal $\ddot{q}_i$ in joint space is not measurable. Cousequently, the acceleration signal $\ddot{q}_i$ is replaced by its estimated value $\hat{\ddot{q}}_i$, which is simply obtained from the differential equation of motion (figure 1):
$$D(q)\hat{\ddot{q}}_i = K_t I_a^{Ref} - \hat{h}(q, \dot{q}) \tag{15}$$

In equation (Eq. 15) $D(q)$ is the main inertia of the robot axes, $K_t I_a^{Ref}$ is the active drive torque developed by the DC motor and is a known quantity, and $\hat{h}(q, \dot{q})$ is the unknown waIue of the load torque. To achive zero error dynamics an asymtotic observer is inserted into the control scheme in figure 1:
$$\frac{dp(t)}{dt} = |\ddot{q} - \hat{\ddot{q}}| = |h(q, \dot{q}) + \hat{h}(q, \dot{q})| \tag{16}$$

It is prescriebed with pole location $s = -g/I_n$, I_n is the nominal inertia of robot link and servomotor.

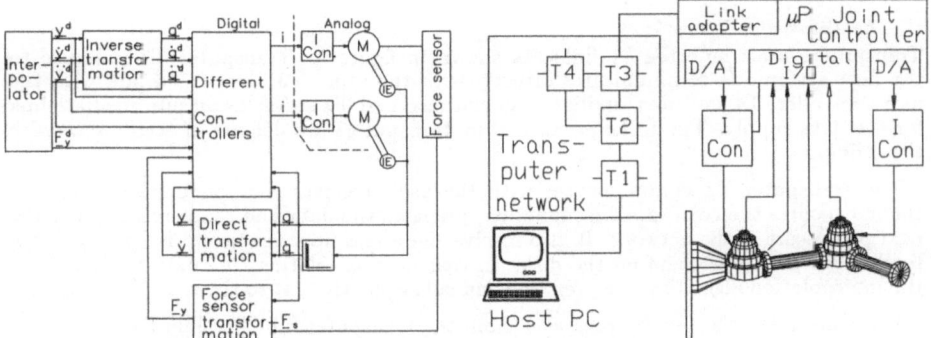

Figure2: Data path in Controller   Figure3: Multiprocessor robot controller

The controller input data are obtained from the interpolator. The interpolation algorithm is based on the $\sin^2$ acceleration profile which enables smooth motion (no jerks). The flow of the necessary data inside the controller is shown in figure 2. The interpolator generates trajectories in the Cartesian space. The values of position, velocity and acceleration are transformed via the inverse robot kinematics into joint angles, angular velocities and angular accelerations. The actual angles and angular velocities are transformed into position and velocity in the Cartesian space by means of direct kinematics. In this way different algorithms can be realized. In our case, however, only the values in the Cartesian space and the calculations of direct kinematics are needed. Inverse transformations are needed only for the remaining algorithms which are implemented in our experimental robot controller.

2. Robot Controller

Our robot controller, presented in figure 3, is likewise transputer based (McKeever, 1992). Its kernel is a card with four T800 transputers. Additional transputers may be added. The card is located in a PC that functions as an interface to the user. The connection with the robot runs via a microprocessor system with a 68010 processor. At present, the controller is connected to a Scara laboratory mechanism with 2 DOF.

Figure 2 shows the basic tasks which are realised in our robot controller. Trajectory planning is generated in Cartesian space. Linear and circle interpolator procedures with the use of the $\sin^2$ profile are realised. Control in joint coordinates needs the inverse kinematics transformation. The inverse transformation is not necessary for the described Cartesian space controller. Direct kinematics transformation and data transformation of the force sensor are also implemented.

These tasks must be distributed on single processors when we use the multiprocessor system. We used static distribution during the software development to achieve shorter executing times and because there was no operating system on our transputer network.

From the definition of tasks we get the following basic processes: planning and coordination of movements (interpreter), interpolation, direct and inverse kinematics, control with transformations, protection (limitation of the axis motion). On a single processor system the tasks are performed consecutively and on a multiprocessor they are distributed among the processors, in which case it may occur that, due to the sequential nature of some processes, only some processors are active while the others are waiting for data. Consequently, it is necessary to distribute the processes. Only in this way all the advantages of a multiprocessor system are exploited.

A. *The PC Level*

The PC is used as a host computer for communications with the user. It enables to enter data and instructions with menu selection, and to store programs and results. All results can be seen as graph on PC screen. We can also represent results on the oscilloscope over the D/A chanels of the joint controller. The robot controller program is executed in the transputer network.

B. *The Transputer Network*

Transputers are connected by links, as shown in figure 3. Transputer T1 is linked to the host system. It communicates directly with the user. T3 serves as a link with the axis controller, T4 executes auxiliary calculations, and T2 serves as an intermediate link for the data supply. The tasks performed by transputers are defined in accordance with these links.

- The transputer T1 communicates with the user, receives the commands, interprets them, executes the corresponding function, prepares the data and sends messages to the next processors in the network. It also receives messages from the network, conveys them to the user or stores them on the disk. In case of interpolation, the data necessary for the interpolation algorithm are prepared and subsequently sent to the second transputer.

- The transputer T2 receives messages from the transputer T1. Possible transient messages are sent to the corresponding transputer (T3 and/or T4). If interpolation is required, the interpolation algorithm is implemented and the values of coordinates are sent forth together with the remaining messages. Messages coming from transputers T3 and T4 need to be transformed so that the processing on T1 will be possible.

- The transputer T3 implements the control algorithm with transformations from figure 1. The desired values are received from T2, whereas the real values are received from the axis controller. One part of the transformation process was parallelized because of the the extent of tasks. At the beginning of computing these data are sent to the transputer T4, and when the process on T4 needs them it takes them. After computing the desired values of motor torques the programmed robot movement limits are tested (T3 and T4 cooperate again). Finally the data are transferred to the axis controller and sampling times are synchronized.

C. *The Axis Controller*

The axis controller is represented by one or several microcomputer systems which are in charge of the operation of one or several axes. They must generate all the signals necessary for each axis and measure the positions, velocities..., etc. A link is used for the connection with the transputer network. The transfer through the link is extremely fast (20 Mbit/s), so bottlenecks should be avoided (e.g. we use a transputer as a processor).

3. Realization

The distribution of processes among transputers and the parallelization of process executions were used. In this way sampling times of $1ms$ were reached, which is several classes better than what the controllers available on the market offer. Robot motion is controlled in the Cartesian space according to acceleration, velocity and path.

The test movement is a circumference in the first quadrant. The interpolator generates the data necessary for the execution of the test movement. The acceleration is made with $\sin^2$ profile with an amplitude of $0.15 m/s^2$, the velocity with a constant velocity phase of 0.1 m/s and the radius of the circumference is 0.03 m.

The deviation of the actual radius from the desired one is given in figure 4. The actual radius exceeds the desired one by 0.15 mm at the most. The deviaton of the radius obtained, when a common position cascade controller is used, is greater: the actual radius exceeds the desired one by 0.5 mm at the most. The static position error is smaller than 0.05 mm in both cases. As regards the continuity of velocity, the controller in the Cartesian space shows very good trajectory tracking, whereas the cascade controller tends to be slower or faster. The results presented so far refer to the measurements performed on the motor side. The position and velocity are measured on the motor axis, so the freeplay and the elasticity of gears and other mechanical components are not considered. Consequently, greater deviations occur in the performed movement. The drawn circumference and the square are shown in figure 5. Very good tracking is observed for movements at the slope of 135 *degrees*, whereas all other directions show greater deviations which are due to mechanical influences of the mechanism. An inverse gear model could be used to eliminate these deviations (Pritschow et. al.,1992), which will be the focus of our future investigations.

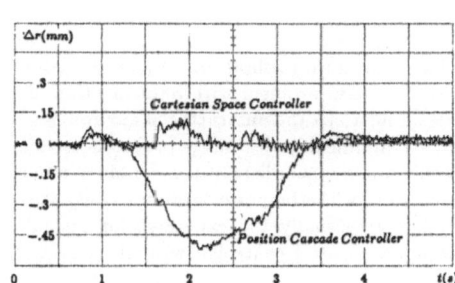 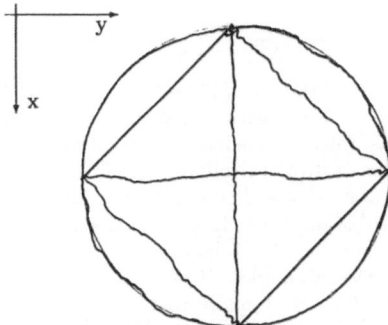

Figure4: The deviation of the radius from the desired circumference

Figure5: The circumference drawn by the Scara mechanism

The obtained results confirm the correctness of our concept. In the system a lot of free time is available. The used microcomputer system hinders a better time efficiency, so the communication between the transputer and the periphery will have to be speeded up. The best results can be reached with memory-maped I/O port (Zhang et. al., 1992). The described and experimentaly tested control algorithm in Cartesian space consist of the acceleration feedback and disturbance torque estimation. It assures a good dynamic performance even in presence of an initial condition mismatch, parameter perturbations and disturbances. However, total insensitivity of the system to disturbances is not possible, but tracking errors are controlled and the dynamic system is asymptotically stable.

References

K. Jezernik, B. Curk, J. Harnik, A. Šabanović, (1992). VSS Control of Robotic Manipulators by Joint Acceleration Controllers, IMACS/SICE 92, Kobe, Japan

S.Komada, K.Ohnishi, (1989). Control of robotic manipulators by joint acceleration controller, Proceedings of the IECON'89, Philadelphia, Pennsylvania, pp. 623-628

J.D.M.McKeever, (1992). Using Transputer in a robot programming and control system, Microprocessing and Microprogramming 34, North-Holland, pp. 117-120

G.Pritschow, H.Klingel, M.Bauder, A.Horn, (1992). Erhoehung der Bahngenauigkeit von Industrieroboter, Robotersysteme 8, pp. 162-170, Springer-Verlag

D. Q. Zhang, C. Cecati, E. Chiricozzi, (1992). Some Practical Issues Of The Transputer Based Real-Time systems, IEEE 0-7803-0582-5/92, pp. 1403-1407

Implementation of a Low Cost Robot Controller PC-ROBOCONT on Hydraulic Robot for Spray Painting G-201

B. Nemec, L. Zlajpah, S. Mrak
Jozef Stefan Institute, Robotics Laboratory
Ljubljana, Slovenia

Abstract: The paper describes the implementation of the low cost robot controller PC-ROBOCONT on hydraulic spray painting robot G-201. PC-ROBOCONT is a low cost robot controller based on popular PC 386. For use with the spray painting robot we added CP teaching and execution capability and conveyor tracking feature to the RRL robot programming language.

1 Introduction

Statistics of the use of the robots (Karlsson, 1992) have shown that robots are employed mainly in large factories with well structured material flow. The main problem with implementing industrial robots in medium and small sized companies seems to be the unstructured environment and the cost of the robotic cell. The unstructured environment poses many problems in implementation of an industrial robots and usually requires more complicated control of an robot and use of many external sensors, which can be done only with the powerful robot controller. The commercially available advanced robot controllers are usually too expensive for a small company. In addition, it often happens that we need very special controller with the features not provided in standard robot controller.

As a contribution to the above mentioned problems we developed low cost advanced robot controller PC-ROBOCONT, based on the popular IBM compatible personal computer AT 386. PC-ROBOCONT is particularly convenient for use in small to medium sized companies, research and development institutions, schools, universities, etc. The main advantage of our robot controller is low cost and high performances at the same time and an open architecture, which allows experimental work in robotics.

2 Controller architecture

Commercially available robot controller are usually based on multiprocessor bus (e.g. VME, MULTIBUS, etc.). They have hierarchical structure with powerful central processor for intensive arithmetic operations and peripherical processors for axis control, various sensors control (e.g. vision sensor) and external communications units (e.g. hand held programming units, link to the external cell computer). Communication between the operator and the machine is provided trough special heavy duty control panels.

PC computer is not intended for industrial use in it's original appearance. Due to the increased popularity of PC a number of extension modules have been developed in the past years, that allow PC to be used as a laboratory measurement computer. On the other hand many manufacturers have build industrial PC-s. This are completely redesigned computer and they have only

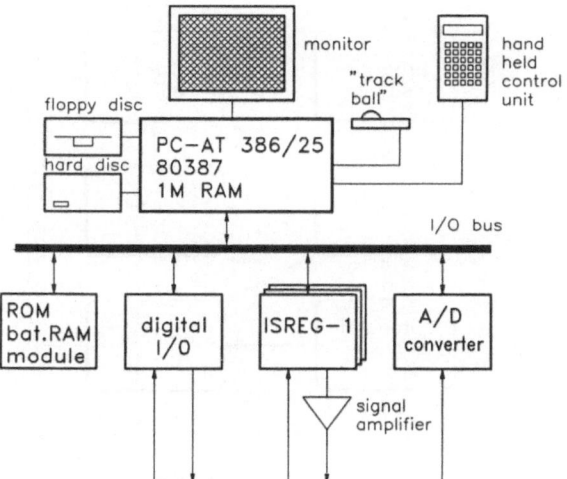

Figure 1: PC-ROBOCONT architecture

program compatibility with the original PC-s, but the cost of such computer is much higher comparing to the ordinary PC.

With PC-ROBOCONT we build an advanced controller using the standard AT-386 compatible computer. The block scheme of the controller is shown on Fig. 1. The basis of the controller forms PC AT 80386 computer with arithmetic coprocessor 80387 (or 80486 based computer) with 1M RAM hard disc, 3.5"floppy drive, VGA monitor and the track ball. For some critical applications we can substitute hard disc and floppy disc drive with ROM - battery backup RAM module. We developed special motion controller ISREG-1 module using National LM628 motion controller IC (National Semiconductor, 1988). Each module can drive two AC or DC servo motors or hydraulic motor and read position and velocity from incremental encoders. Motion controller has build in discrete PID control law with a number of monitoring and safety functions. It uses 32 bit position counters and is thus also convenient for most precise NC machines. Various sensors are connected to the controller by optically protected digital and analog input-output modules. The control panel of the controller is substituted with VGA monitor and the track ball. Using special graphic interface developed at our institute, the control panel with all necessary buttons, keyboard, potentiometers, warning lights, displays, etc. is simply drawn on VGA monitor. The keyboard and potentiometers are commanded by moving track ball. If the application requires additional lights or buttons, they are drawn on the display with minor modification of the software.

3   Robot programming language

In PC-ROBOCONT we implemented RRL (Riko Robot Language), originally developed in our institute for RIKO 106 welding robots and STEFAN VME based robot controller (Nemec, 90; Nemec, 1992). The block scheme of RRL program environment is presented on Fig. 2 .

The RRL is composed of the following program modules: The **RRL kernel** contains interpolation procedures and kinematic transformations. Smooth trajectory generation is achieved trough 4-1-4 order polynomial splines (Paul, 1981). Among the interpolation we can choose between straight line, circle, smooth transition between intermediate points and polynomial splines trough a set of points. Kinematic transformations contain solution for 6 joint

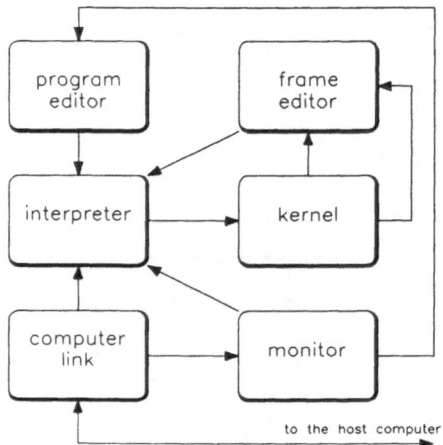

Figure 2: Structure of the RRL environment

articulated robot, 6 joint cartesian and 4 joint Scara robot. Up to 9 robot axes can be controlled. The remaining three axes are considered as redundant robot axes thus allowing the synchronized move of the robot with the external positioning device. **RRL editor** is menu driven editor that allow simple editing of textual robot program with syntax check. **RRL interpreter** module executes commands of special high level robot program language. There are the following groups of commands: interpolation commands, frames definition commands, arithmetic commands for integer, real and vector data, program control commands and commands for interaction with the environment. We included also commands for the arc-welding process, force control using 6-dof force sensor and weld seam tracking. **RRL frames editor** module is used for the definition of the required points (with points we mean position and orientation of the robot and the position of addition axes of an external position device. This generalized point is called frame) by teach-in or by entering their numerical value. **RRL computer** link is used for program transfer between an external computer and for the connection of the controller to the master cell computer. All RRL modules are connected by **RRL monitor**, which can select the desired operation and allows execution of the program in various modes : normal, step-by-step, interrupt mode, reverse mode, repeat mode, etc.

3.1 Control of hydraulic motors

Spray painting robot G-201 is driven by hydraulic motors. (Fig. 3). The output of the DAC converter from control module is amplified and controls directly the hydraulic valves. The implemented control algorithm is PI. Integration part improves steady state error, but may cause instability of the system and therefore only a small amount of the integration part is acceptable. The tracking error is diminished by velocity and acceleration feed-forward signal. Unfortunately, LM628 motion controller has no compensation input. We solved this problem by modified feed-forward scheme presented on Fig. 4. The inverse controller transfer function $\frac{1}{G_r}$ is realized with software and the output of the modified feed-forward control is added to the reference position signal. This solution is not completely equivalent of the hardware compensation input, while the gain of the controller G_r in modified scheme diminishes the

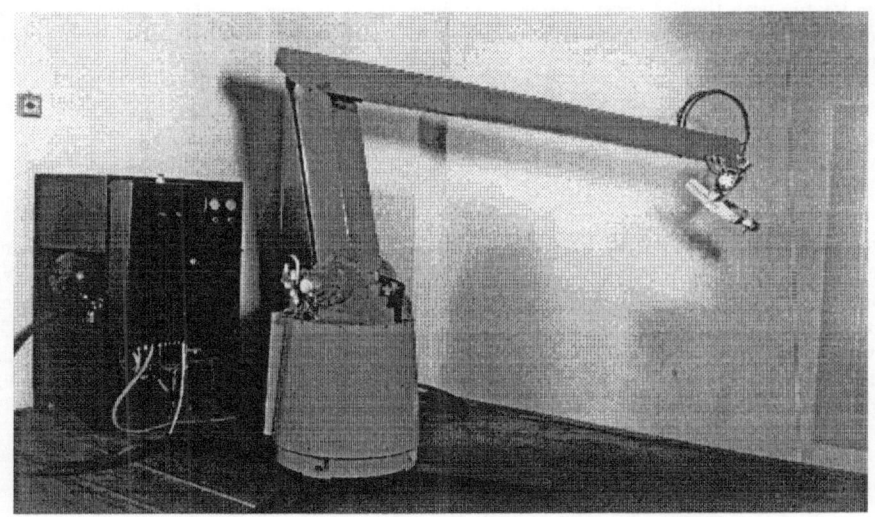

Figure 3: Spray-painting robot G-201

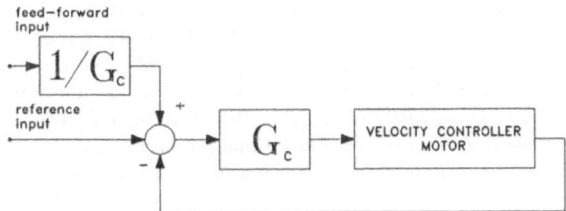

Figure 4: modified feed-forward control scheme

resolution of the compensation signal.

3.2 Trajectories for the spray painting robot

Original RRL program package had no possibility of definition of an arbitrary trajectory. The only way to define the trajectory was to connect the lines and arcs or by polynomial splines, which is not convenient for the spray painting. The most efficient and simple way to define an arbitrary trajectory is manual guidance of the robot. During the manual guidance the controller records the world coordinates of the robot with the required sampling rate. Storing such a large amount of data would quickly exceed the memory capacity of the controller. Therefore, an appropriate data reduction algorithm must be used. The most popular reduction algorithm is picewise linear modelling (Libre, 1982). In this algorithm we approximate the trajectory with straight line, as long as the set of successive data stays in the prescribed tube around the approximated line. The drawback of the above algorithm is that the velocity is discontinuous at the joint points of the approximated lines, which can induce natural frequency of the mechanism. To avoid this problem we developed another algorithm, where we sample the trajectory with the reduced sampling frequency (10 Hz). During the execution, the sampled frequency is interpolated with the 3rd order interpolation filter. That assures that the accelerations of the resulting trajectory are continuous. The interpolation algorithm is described by the following equation:

$$p(t) = a + bt + ct^2 + dt^3$$

$$a = \frac{p(k-1) - 4p(k-2)}{+}p(k-3)6 \qquad b = \frac{p(k-1) - p(k-3)}{2N}$$

$$c = \frac{p(k-1) - 2p(k-2) + p(k-3)}{2N^2} \qquad d = \frac{p(k) - 3p(k-1) + 3p(k-2) - p(k-3)}{6N^3}$$

where N is the number of the interpolation points, $p(k)$ is the cartesian position at the time sample k, $p(t)$ in the interpolated position and t is the interpolation counter, $t = \{1...N\}$. The resulting trajectory is then feed to the control algorithm with time interval 0.01 s, where is additionally interpolated using 2nd order interpolation with 5 intermediate points. The advantage of the described algorithm is smooth response, the disadvantage is that is keep storing data while the robot stands still. To avoid this we can combine our 3rd order interpolation algorithm with straight line interpolation algorithm.

4   CONCLUSIONS

In the paper we presented low-cost robot controller PC-ROBOCONT. With 25 MHz processor 80386-387 we achieve 100 Hz sampling rate in trajectory generation including kinematical transformations. The performance is comparable with our VME based controller STEFAN 109. Due to the robust design we expect that the VME based controller is more reliable in heavy industrial environments. On the other hand the PC-ROBOCONT is more than five times cheaper comparing to the VME based controller. PC-ROBOCONT is therefore very convenient for the applications, where we need high performances and low cost at the same time and the reliability is not of the essential importance. PC-ROBOCONT is also very convenient for the education, research and development because it allows easy experimentation and modification of build-in algorithms. It is designed as a completely open system and the software is entirely written in Turbo Pascal.

References

Karlsson, J.(1992). Yearbook on Industrial robot statistics 1991. International Federation on Robots

Nemec, B. and Ružič. A and Ilc. V. (1992). RRL - robot programming environment. Informatica, Vol 16, No 1

LM628/LM629 Precision Motion Controller (1988). National Semiconductor Corporation

Paul, R.P. (1981). Robot Manipulators: Mathematics,Programming and Control. MIT Press, Massachusetts

Libre, M. and Bel, G. (1982). Data compression method for the recording of industrial robot trajectories. Proceedings of 12th International Symposium on Industrial Robots, Paris

Robust Adaptive Lyapunov-Based Robot Control

V. Vesely, D. Kalas
Faculty of Electrical Engineering
Slovak Technical University, Bratislava

Abstract: Robust adaptive control scheme is presented for robotic manipulators. The effects of parametric and dynamic uncertainties in the robot mathematical model are suppressed by adaptive control scheme which ensures that the stability measure of nominal system will increase in time until there are no transient processes.

1 Introduction and problem statement

High - performance robotic manipulators applications require sophisticated controller design in order to guarantee excellent tracking of fast desired motions. This paper represents the application of globally asymptotically stable adaptive control philosophy capable of handling MIMO nonlinear systems with uncertain parameters to robotic manipulator controller design. The tracking performance is uniformly good despite severe abrupt parametric and desired trajectory step changes as has been confirmed by simulations. Consider a nonlinear time - varying dynamic system

$$\underline{\dot{x}} = \underline{f}(\underline{x}, t) + \underline{b}(\underline{x}, \underline{u}, t) \tag{1}$$

where
$\underline{x} \in \mathcal{R}^n$; $\underline{u} \in \mathcal{R}^m$ are state and plant input variables.
The design problem of the control agent is twofold: to achieve closed - loop stability and to reduce the influence of the dynamic and parametric uncertainties on the system (1) with proposed adaptive control law which is given by the following formula

$$\underline{u} = \underline{q}(\underline{x}, \underline{r}, t) \tag{2}$$

$$\underline{\dot{r}} = \underline{g}(\underline{x}, \underline{r}, t)$$

where $\underline{r} \in \mathcal{R}^p$ denotes the controller parameters under the adaptation.

2   Dynamic system stabilization problem

Define a function $V : \mathcal{R}^n \times J \to \mathcal{R}_+$ as a Lyapunov function $V(\underline{x}, t)$ for the system (1) with $\underline{u} = \underline{0}$, where $J = \{t : t \in\, <t_0, \infty)\}$.

Definition 1 *The measure of stability of dynamic system (1) with the control law (2) is given by the following equation*

$$\alpha \, (\underline{x}, \underline{r}, t) = -\frac{\frac{dV(\underline{x}, \underline{r}, t)}{dt}}{V(\underline{x}, t)} \qquad (3)$$

Definition 2 *Consider a nonlinear time - varying system of the form (1). We say that the system (1) is P - stabilizable if there exist both a Lyapunov function $V(\underline{x}, \underline{r}, t)$ and a control algorithm (2) that on the set $\mathcal{R}^n \times J \in \mathcal{R}^n$ the following condition is valid*

$$\frac{dV}{dt} = \frac{\partial V}{\partial t} + (gradV)^T[\underline{f}(\underline{x}, t) + \underline{b}(\underline{x}, \underline{q}(\underline{x}, \underline{r}^*, t), t)] \leq 0 \qquad (4)$$

In order to ensure the stability of the investigated system, the stability measure $\alpha\,(\underline{x}, \underline{r}, t)$ has to be increased when the system (1) with (2) is not stable and/or desired control performance is not met. Consider the Lyapunov function with the adaptive controller (AC)

$$V_a = V + (\underline{r} - \underline{r}^*)^T(\underline{r} - \underline{r}^*) \qquad (5)$$

From the condition for the change of the stability measure of the system (1) with (2) as a controller parameters $\underline{r}$, one has the following AC algorithm (Veselý, 1993)

$$\dot{\underline{r}} = -\beta\frac{\partial}{\partial \underline{r}}[(gradV)^T\underline{b}(.)] \qquad (6)$$

with β constant. Assume that instead of (6) one may use one of the following algorithms

$$\dot{\underline{r}} = \underline{\xi}(\underline{x}, \underline{r}, t) \qquad (7)$$

$$\underline{r} = \underline{\mu}(\underline{x}, t)$$

where $\underline{\xi}(.)$ and $\underline{\mu}(.)$ are suitably selected vector valued functions.
The proof of the sufficient stability conditions of the system (1) with (2) and (7) are given in (Veselý, 1993).

3 Adaptive robot control, simulations and conclusion

Consider the dynamics of an n - joint rigid - link robotic manipulator described by well-known differential equation (Paden and Sastry, 1987)

$$M(\underline{q})\ddot{\underline{q}} + \underline{c}(\underline{q}, \dot{\underline{q}}) + \underline{g}(\underline{q}) = B\underline{u}(.) \qquad (8)$$

Suppose that the following property is valid for (8) (Paden and Sastry, 1987).
Property *Let $[\underline{q}_d ; \dot{\underline{q}}_d]^T \in C^1$. Then the trajectory tracking error is given by the following equation*

$$[\underline{e} ; \dot{\underline{e}}]^T = [\underline{q} ; \dot{\underline{q}}]^T - [\underline{q}_d ; \dot{\underline{q}}_d]^T \qquad (9)$$

The tracking problem is the following: the designer should find a feedback control law (2) such, that for any initial state of the system (8)

$$\lim_{t \to \infty} [\underline{e}; \dot{\underline{e}}]^T = \underline{0} \qquad (10)$$

holds.
Choose $D \in \mathcal{R}^{n \times n}$ such that $D > 0$. Define the new vector

$$\underline{s} = [D \mid I][\underline{e}; \dot{\underline{e}}]^T \qquad (11)$$

and the control algorithm (2) with the adaptation

$$\underline{u}(.) = -rB^{-1}K\underline{s} \qquad (12)$$

$$\dot{r} = \beta \, \underline{s}^T \underline{s} \qquad (13)$$

where $K = diag\{k_{ii}\}$; $K \in \mathcal{R}^{n \times n}$; $k_{ii} > 0$. The proposed AC (12), (13) ensures that the stability measure of the control system (8) will increase in time if $\dot{r}$ is not identically equal to zero.
Consider a Lyapunov function for the system (8), (9), (11) given in the form (5). For the time derivative of V_a along the solution of (8), (9), (11), (12) and (13) one obtains

$$\frac{dV_a}{dt} = \underline{s}^T \dot{\underline{s}} + \dot{r}(r - r^*) =$$

$$= -\underline{s}^T(rM^{-1}K + \beta(r^* - r))\underline{s} + \underline{s}^T[D\dot{\underline{e}} - M^{-1}(\underline{c} + \underline{g}) - \ddot{\underline{q}}_d] \leq \qquad (14)$$

$$\leq -r\lambda_{min}(M^{-1}K) \parallel \underline{s} \parallel^2 + \parallel \underline{s} \parallel [\parallel D\underline{\dot{e}} \parallel + \parallel M^{-1}(\underline{c}+\underline{g})+ \parallel \underline{\ddot{q}}_d \parallel]$$

where λ_{min} denotes the minimum eigenvalue of a matrix, and M is positive definite.

Theorem 1 *Let the manipulator dynamics and the adaptive controller be described by (8), (12) and (13). Under the assumption* **Property***, there exist such values of matrix K, that the investigated system is stable and asymptotically stable with respect to variable $\underline{q}$.*

Proof
Due to the **Property**, the second part of (14) is locally bounded. There exist positive numbers r and $k_{ii} > 0$; $i = 1, \ldots, n$ such that

$$\frac{dV_a}{dt} < -\varepsilon(\parallel \underline{z} \parallel) < 0 \tag{15}$$

where $\varepsilon(.)$ is a continuous and increasing for every $\parallel \underline{z} \parallel$ and $\varepsilon(0) = 0$ positive real function of $\underline{z}^T = [\underline{s}^T ; r - r^*]$.
Because of (15) the system is stable and asymptotically stable with respect to variable $\underline{s}$, i.e.

$$\lim_{t \to \infty} \underline{s} = \underline{0}$$

and from (11) one has

$$\underline{\dot{e}} = -D\underline{e} \tag{16}$$

Owing to $D > 0$ the system (16) is asymptotically stable with respect to $\underline{e}$ which proves the sufficient stability conditions of the system (8), (12) and (13). □

A two - link three degrees of freedom planar robotic manipulator was considered. The first link is free to revolve in the vertical plane (revolutionary joint-variable q_2) while its legth can vary (prismatic joint-variable q_1) and at its end there is the second link free to move in the same plane (revolutionary joint-variable q_3). For the simulation purposes the diagonal matrix $D = diag\{d_{ii}\}$; $d_{11} = 20$; $d_{22} = d_{33} = 100$ was chosen. The entries of the matrix K were $k_{11} = 30$; $k_{22} = k_{33} = 100$ and β was $\beta = 25$.
The trajectories to be tracked are in all degrees of freedom given in the form of sinusoidal functions $q_{di} = 0.5sin(0.4t) + 0.5$; $i = 1, 2, 3$.

In Fig. 1 there are tracking errors while in $t = 5sec$ the abrupt manipulator severe payload change took place from $2kg$ to $20kg$ - the effect of this disturbance is negligible - and in $t = 8sec$ there was a unit step change in the desired state trajectory for the second degree of freedom q_{d2} from $0.5sin(0.4t) + 0.5$ to $-0.5sin(0.4t) + 0.5$. The control philosophy suggested in this paper exhibits insensitivity to plant parameters uncertainty and variations and enjoys the global asymptotic stability.

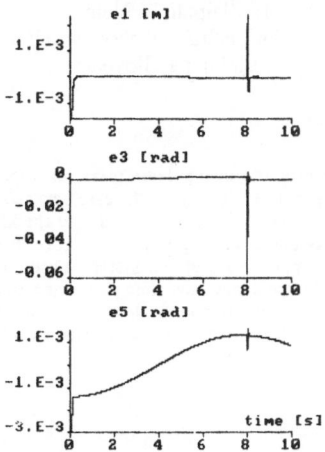

Fig. 1 Tracking errors - parameter step change in $t = 5sec$ and severe step change in preplanned trajectory in $t = 8sec$.

4   References

PADEN, B. E., SASTRY, S. S. (1987). A Calculus for Computing Filippov's Differential Inclusion with Application to the Variable Structure Control of Robot Manipulators, *I.E.E.E. Transactions on Circuits and Systems*, **CAS - 34**, 73 - 82.

VESELÝ, V. (1993). Large Scale Dynamic System Stabilization Using the Principle of Dominant Subsystems, *Kybernetika*, **29**, 48 - 62.

Stability of Hybrid Position/Force Control Scheme During Free Motion

L. Zlajpah, B. Nemec
Jozef Stefan Institute, Robotics Laboratory
Ljubljana, Slovenia

Abstract: The kinematics structure of the manipulator plays an important role concerning the stability problems of the hybrid position/force control. Namely, in the case when some axes are position controlled and others are force controlled, the system may become unstable due to the kinematic structure. The paper deals with this problem in case when the system is moving in the free space. It is shown that the instability region cannot be located only by kinematic parameters of the system. Furthermore, the size and the location of instability regions depend as well on kinematic and dynamic parameters of the system as on the chosen control scheme and controller parameters. Another important result is, that if the system is unstable for one configuration in particular Cartesian position, then there exists a stable configuration for the same position.

1 Introduction

Position and force control are required if the robot has to perform advanced tasks, i.e., the motion is constrained by the task environment. One of the control methods used in such systems is a hybrid position/force control proposed by Raibert and Craig (1981). The method has been further developed (Khatib, 1987; Zhang and Paul, 1985).

When the hybrid control is applied, the stability of the system must be guaranteed during all the phases: free motion, contact and constraint motion (the end–effector is in the contact with the environment). Concerning these general problems of hybrid position/force control most of the research has been focused on the contact instability and on the stability after the contact (Whitney, 1987; Hogan, 1988; An and Hollerbach, 1989; Kazerooni, 1990; Stokič, 1991; Yabuta, 1992). The main issue in these stability analyses is the dynamics of the system. On the other hand, during free motion another type of instability plays an important role. An and Hollerbach (1989) referred to it as *kinematic instability*. Namely, in case when some axes are position controlled and other are force controlled (the task is usually defined in Cartesian space), the system may become unstable due to the kinematic structure. The problem was further analysed by several authors. Zhang (1989) tries to find the necessary and sufficient conditions under which the hybrid control is stable due to kinematic configuration. However, he uses only the kinematic model and ignores the dynamics model and controller structure and parameters. The detailed stability analyses (using Lyapunov's method) of hybrid control scheme considering different factors are given in (Yabuta, 1992), but the presented fundamental stability conditions are hard to implement in practice.

This paper deals with the problem of the kinematic instability in case when the system is moving in the free space, i.e., contact forces are zero. The objective is to give the insight into the problem and to propose how to overcome this problem in practice. Following the idea presented in (Zhang, 1989), we try to find the regions where the hybrid control as proposed by Raibert and Craig (1981) and Zhang and Paul (1985) is unstable due to the kinematic configuration.

It is shown that the instability region cannot be located only by kinematic parameters of the system. Furthermore, the size and the location of instability regions depend as well as on kinematic and dynamic parameters of the system as on the chosen control scheme and controller parameters. Another important result is, that if the system is unstable in particular Cartesian position for one configuration (i.e., solution of the inverse kinematics), then there exists a stable configuration for the same position. So, there is a possibility to avoid the instability region by choosing the proper configuration. To give the better insight into the problem, an investigation is carried out through theoretical analysis and simulation on a planar robot with two rotational axes.

2 Hybrid control scheme

The main characteristic of the hybrid position/force control scheme is in the division of the system into position- and force-controlled subsystems. Although the separation may be done in any space, it is reasonable to make it in Cartesian (or task) space. A selection matrix $\mathbf{S}$ is defined to separate the position-controlled and the force-controlled subsystems (Raibert and Craig, 1981). Zhang and Paul (1985) have modified the scheme. Namely, they have used Jacobian matrix instead of the inverse kinematic transformation in the feedback loop. The control law is in the form:

$$\mathbf{u} = \mathbf{K}_p \mathbf{J}^{-1} \bar{\mathbf{S}} \mathbf{J} (\mathbf{q}_d - \mathbf{q}) + \mathbf{K}_v \mathbf{J}^{-1} \bar{\mathbf{S}} \mathbf{J} (\dot{\mathbf{q}}_d - \dot{\mathbf{q}}) + \mathbf{K}_f \mathbf{J}^T \mathbf{S} (\mathbf{F}_d - \mathbf{F}) \tag{1}$$

As we can see, the errors are calculated in joint space and then transformed via Jacobian matrix to the Cartesian space. This transformation is only the approximation for transforming differential motion (Paul, 1981). Therefore, it is valid only for small errors.

An and Hollerbach (1989) claim that such control methods may be unstable and Zhang (1989) shows that this happens in certain manipulator configurations using revolute joints. In his analysis of instability regions Zhang ignores the influence of the dynamic parameters. Therefore his results are not correct. Our goals are: first to define the size and the location of instability regions based on the kinematic and dynamic parameters of the system, on the chosen control scheme and the controller parameters; second to define a supremum of the instability regions by using only kinematics parameters; third to propose a minor improvement to stabilize the control law 1.

To present the essential effects we consider only the motion in free space. Then, the contact forces are zero, $\mathbf{F} = 0$ and without loss of generality we can choose also $\mathbf{F}_d = 0$. This phase is a transition phase when the robot is approaching the contact surface. Additionally, if we consider the set-point control ($\dot{\mathbf{x}}_d = 0$), the control law 1 can be simplified and the following law is obtained:

$$\mathbf{u} = \mathbf{K}_p \mathbf{J}^{-1} \bar{\mathbf{S}} \mathbf{J} (\mathbf{q}_d - \mathbf{q}) - \mathbf{K}_v \mathbf{J}^{-1} \bar{\mathbf{S}} \mathbf{J} \dot{\mathbf{q}} \tag{2}$$

3 Stability analyses

In (Yabuta, 1992) the fundamental stability conditions of the nonlinear hybrid position/force control system are given by using the Lyapunov's method. He shows that the conditions to prove the stability of the system can not be satisfied due to $\mathbf{J}^{-1}$ term effect and concludes that Raibert and Craig's control scheme is unstable, but his results are considering only the global stability of the system. Using another approach we will show that there are only some regions (in practice even very small) where the system becomes unstable.

With respect to n joint coordinates $\mathbf{q}$, the motion of n D.O.F. manipulator can be given in a well known form:

$$\mathbf{H}(\mathbf{q})\ddot{\mathbf{q}} + \mathbf{h}(\mathbf{q},\dot{\mathbf{q}}) + \mathbf{g}(\mathbf{q}) = \tau \tag{3}$$

where $\mathbf{h}(\mathbf{q},\dot{\mathbf{q}})$, $\mathbf{g}(\mathbf{q})$, and τ represent n-dimensional vectors of Coriolis and centrifugal, gravity, and joint driving forces, respectively, and $\mathbf{H}(\mathbf{q})$ is $n \times n$ inertial matrix. The gravity forces are supposed to be compensated. Assuming small velocities $\dot{\mathbf{q}} \approx 0$ (Coriolis and centrifugal forces are negligible) and linearizing the Eq. 3 about some working position $\mathbf{q}$ the following model is obtained:

$$\tau = \mathbf{H}(\mathbf{q})\ddot{\mathbf{q}} \tag{4}$$

Defining state variables $[\mathbf{q}^T, \dot{\mathbf{q}}^T]$ the system is described in state space as (for the simplicity the dependancy on working point $\mathbf{q}$ is omitted):

$$
\begin{aligned}
\begin{bmatrix} \dot{\mathbf{q}} \\ \ddot{\mathbf{q}} \end{bmatrix} &= \begin{bmatrix} \mathbf{0} & \mathbf{I} \\ -\mathbf{H}^{-1}\mathbf{K}_p\mathbf{J}^{-1}\bar{\mathbf{S}}\mathbf{J} & -\mathbf{H}^{-1}\mathbf{K}_v\mathbf{J}^{-1}\bar{\mathbf{S}}\mathbf{J} \end{bmatrix} \begin{bmatrix} \mathbf{q} \\ \dot{\mathbf{q}} \end{bmatrix} + \begin{bmatrix} \mathbf{0} \\ -\mathbf{H}^{-1}\mathbf{K}_p\mathbf{J}^{-1}\bar{\mathbf{S}}\mathbf{J} \end{bmatrix} \begin{bmatrix} \mathbf{q}_d \\ \dot{\mathbf{q}}_d \end{bmatrix} \\
&= \mathbf{A} \begin{bmatrix} \mathbf{q} \\ \dot{\mathbf{q}} \end{bmatrix} + \mathbf{B} \begin{bmatrix} \mathbf{q}_d \\ \dot{\mathbf{q}}_d \end{bmatrix}
\end{aligned} \tag{5}
$$

This simplified model is sufficient to study the local stability of the system and to define unstable regions.

The characteristic polynomial $c(\lambda)$ of the system described by the state space model 5 is

$$
\begin{aligned}
c(\lambda) &= |\lambda^2\mathbf{I} + \lambda\mathbf{H}^{-1}\mathbf{K}_v\mathbf{J}^{-1}\bar{\mathbf{S}}\mathbf{J} + \mathbf{H}^{-1}\mathbf{K}_p\mathbf{J}^{-1}\bar{\mathbf{S}}\mathbf{J}| \tag{6} \\
&= |\mathbf{H}^{-1}||\lambda^2\mathbf{H} + \lambda\mathbf{K}_v\mathbf{J}^{-1}\bar{\mathbf{S}}\mathbf{J} + \mathbf{K}_p\mathbf{J}^{-1}\bar{\mathbf{S}}\mathbf{J}| \tag{7}
\end{aligned}
$$

A sufficient condition for the system to be stable is that all poles of the system matrix $\mathbf{A}$ or the roots of $c(\lambda)$ have non-positive real parts. Another method is Routh criterion, i.e., to check the signs of all coefficients in the characteristic polynomial. A necessary condition for the stability is that after any zero root has been removed from the polynomial all the remaining coefficients are present and of the same sign. For a second order polynomial this is also a sufficient condition.

4 A case study

For a case study we have used the model given in (An and Hollerbach, 1989) with the same model and controller parameters (Case 2).

To study the stability, we first evaluate the coefficients of the characteristic polynomial of the system matrix $\mathbf{A}$ in 5. We have selected position control in y-direction and force control in x-direction. So, $\bar{\mathbf{S}} = \text{diag}(0,1)$, and two poles are at the origin. The polynomial is then given by

$$c(\lambda) = \lambda^2(a_4\lambda^2 + a_3\lambda + a_2) \tag{8}$$

Using the following notation

$$\mathbf{J}^{-1}\bar{\mathbf{S}}\mathbf{J} = \begin{bmatrix} J_{11} & J_{12} \\ J_{21} & J_{22} \end{bmatrix} \qquad \mathbf{H} = \begin{bmatrix} H_{11} & H_{12} \\ H_{12} & H_{22} \end{bmatrix} \tag{9}$$

we obtain:

$$a_4 = 1$$

$$a_3 = \frac{J_{22}k_{v,2}H_{11} - J_{12}k_{v,1}H_{12} - J_{21}k_{v,2}H_{12} + J_{11}k_{v,1}H_{22}}{H_{11}H_{22} - H_{12}^2}$$ (10)

$$a_2 = \frac{J_{22}k_{p,2}H_{11} - J_{12}k_{p,1}H_{12} - J_{21}k_{p,2}H_{12} + J_{11}k_{p,1}H_{22}}{H_{11}H_{22} - H_{12}^2}$$

To guarantee the local stability at the equilibrium point, the roots of $c(\lambda)$ must have positive real parts. In order to have a better insight into the close loop dynamics, we calculate additionally the dominant damping ζ of the system. For the comparison we have included the results obtained by the method described in (Zhang, 1989), where the unstable regions are determined only with the matrix $J^{-1}\bar{S}J$. Instead of the element J_{11} (as used by Zhang) we use the element J_{22}. Fig. 1 shows the damping of the system in joint and Cartesian space. It is obvious that

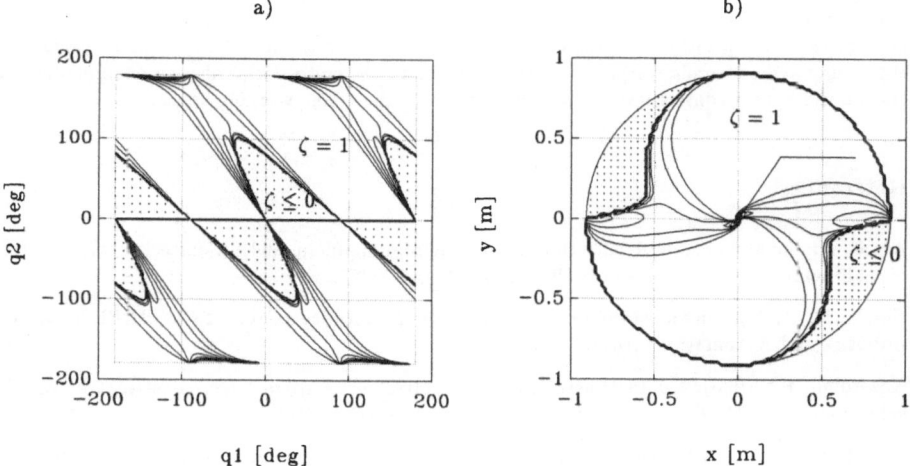

Figure 1: a) Unstable regions in joint space — $\zeta \leq 0$ (step between two lines is 0.2)
b) Unstable regions in Cartesian space for $q_2 < 0$

the unstable regions must be calculated by using complete model and not only the kinematic configuration.

The unstable regions in Cartesian space are similar for $q_2 > 0$ and $q_2 < 0$. The contour plot is only mirrored around x-axis (or y-axis) and the unstable regions for the particular kinematic configuration do not overlap. This means that if the system is in one Cartesian position for one set of $\mathbf{q}$ unstable, then it is for the other set of $\mathbf{q}$ stable.

At the end we want to issue some guidelines how to overcome this problem in practice. Using the information from Fig. 1b) we can place the manipulator and other objects so, that the contact tasks lie in the stable region of the working space. Another possibility is to change the controller parameters or the control scheme. An and Hollerbach (1989) have tried to define certain relationships between $\mathbf{K}_p$ and $\mathbf{K}_v$ to stabilize the system in critical regions. But this approach has no practical meaning because the gains are bounded to other system parameters (e.g. masses, damping, ...). Other authors have proposed different solutions to this problem. A promising change is to add damping to the force controlled subsystem. This can be done

132

by adding the force derivative $\dot{\mathbf{F}}$ to the control law 1 or even better to have position derivative term to be efective for all subsystems. Now the position part of control law is in the form:

$$\mathbf{u} = \mathbf{K}_p \mathbf{J}^{-1}\bar{\mathbf{S}}\mathbf{J}(\mathbf{q}_d - \mathbf{q}) + \mathbf{K}_v(\dot{\mathbf{q}}_d - \dot{\mathbf{q}}) \tag{11}$$

Analyzing the dominant damping of the system with the control law 11, we can see, the modified system has no unstable regions.

5   Conclusion

The paper deals with the problem of the kinematic instability of hybrid position/force control. We have shown that the occurance of the instability is not due to kinematic configuration only, but depends also on dynamic and control parameters. For example, an investigation is carried out through theoretical analysis and simulation on a planar robot with two rotational axes. Another important result is, that if the system is unstable in particular Cartesian position for one configuration (i.e., solution of the inverse kinematics), then there exists a stable configuration for the same position. So, there is a possibility to avoid the instability region by choosing the proper system configuration for a certain task. In the end we give also an example how to change the hybrid control law to make it stable over the whole working space.

References

An, C. H. and Hollerbach, J. M. (1989). The role of dynamic models in cartesian force control of manipulators. Int. J. of Robotic Research, 8(4):51 – 72.

Hogan, N. (1988). On the stability of manipulators performing contact tasks. IEEE Trans. on Robotics and Automation, 4(6):677 – 686.

Kazerooni, H. (1990). Contact instability of the direct drive robot when constraint by a rigid environment. IEEE Trans. on Automati Control, 35(6):700 – 714.

Khatib, O. (1987). A unified approach for motion and force control of robot manipulators:the operational space formulation. IEEE Trans. on Robotics and Automation, 3(1):43 – 53.

Paul, R. P. (1981). Robot manipulators: mathematics, programing and control. MIT Press.

Raibert, M. H. and Craig, J. J. (1981). Hybrid position/force control of manipulators. Trans. of ASME J. of Dynamic Systems, Measurement, and Control, 102:126 – 133.

Stokič, D. M. (1991). Constraint motion of manipulation robots — a contribution. Robotica, 9:157 – 163.

Whitney, D. E. (1987). Historical perspective and state of the art in robot force control. Int. J. of Robotic Research, 6(1):3 – 14.

Yabuta, T. (1992). Nonlinear basic stability concept of the hybrid position/force control sceme for robot manipulators. IEEE Trans. on Robotics and Automation, 8(5):663 – 670.

Zhang, H. (1989). Kinematic stability of robot manipulators under force control. In Proc. IEEE Conf. Robotics and Automation, 80 – 85, Scottdale, AZ.

Zhang, H. and Paul, R. P. (1985). Hybrid control of robot manipulators. In Proc. IEEE Conf. Robotics and Automation, 602 – 607, St. Louis.

Recognition of Three-Dimensional Objects Using Two-Dimensional Subspaces Gained by Intersections

P. Stöhr
Institute of Realtime Systems and Robotics
Technical University, Munich, Federal Republic of Germany

Abstract: This paper presents preliminary considerations on a model-based object recognition system for three-dimensional objects with automatic generation of the models.
Due to the fact that these models are generated by algorithms and do not look like CAD models, the well-known restrictions of standard model-based object recognition systems do not apply to this method.

1   State of the art

Computer vision in general and object recognition using optical sensor systems in particular are rapidly developing fields of research. Using special hardware for the recognition task and very simple environments, robots can work guided by sensors. But normally it is not possible to apply these assumptions, since we have to use normal, and therefore cheaper, hardware. Moreover, even the working field of an assembly robot is not simple enough to enable us to use standard techniques such as object recognition based on the detection of simple geometric elements or on using texture elements. This is the reason why the task of object recognition is far from having been solved in real situations.

1.1   Requirements on an object recognition system used in autonomous robots

An object recognition system designed for autonomous assembly robots has to fulfill several requirements. One is, that the objects must be recognized in almost real time. Furthermore, the system must be able to learn and then recognize every learned object, it is not enough that the system can recognize only a small group of objects, for example polyhedrons.

As the object recognition system is only used in autonomous assembly robots, it is possible to make some assumptions. For example, the way the object is lit does not differ very much between the time of learning and that of recognition and the way the sensor system looks at the scene is known. The number of objects to be recognized is not very large, an robot working autonomously deals with approximately 20 to 30 objects.

Because of the problems which occur when using standard picture processing techniques, or even combinations of them (Schrott, 1993), this paper describes preliminary considerations on an alternative method of model–based object recognition.

2   A new approach

Any given gray–scale picture of a certain scene can be interpreted as a three–dimensional function. After giving all background pixels the value 0, the objects in the scene are those parts with non–zero values. These parts of the picture are the three–dimensional representation of the real objects in the scene. Unfortunately, these three–dimensional representations are too complicated to be used as recognition patterns and have therefore to be simplified.

2.1 Basic idea

The method of simplifying these representations described in this paper uses a simple paradigm of computer graphics which says:

> To solve the hidden–point problem of a given point and a given scene, try to use as much as possible faster and simpler algorithms with a two–dimensional projection. Only if these algorithms fail, should you use the more complex and slower algorithms with a three–dimensional working space.

Applying this idea on the recognition of objects based on optical sensor systems means:

> To recognize a three–dimensional object we have to use two–dimensional characteristics which are simple to find, fast to compute and easy to distinguish.

A two–dimensional characteristic is chosen because a pattern–matching process for two–dimensional data is much faster than that with three–dimensional information.

2.2 Requirements on the two–dimensional characteristic

As described above, the picture given by the sensor system is divided into several small segments each containing one object. This is why transposing the object in the x–y plane must not change the characteristic. The position of the object can be found by the algorithm which divides the original picture.

If the object is rotated around the z axis by the angle α, this angle must influence the characteristic analytically. This angle is needed to describe the orientation of the object and to determine grip positions.

Because the characteristic is used to recognize the object it should be unique for each object. This information is stored, together with the gray–scale picture and a test object T, in a recognition database. During the recognition task, the information stored in this database is used to determine whether an object seen had been learned (and if so, as what) or if it is still unknown.

As the two–dimensional characteristic must be stable in spite of distortions in the data obtained by the sensor system, the algorithm should not rely on the detection of geometric elements.

2.3 Obtaining the two–dimensional characteristic

2.3.1 Evaluation

Generating the two–dimensional characteristic is done in three steps:

- Picture pre–processing step:
  After applying some simple filter operation on the gray–scale picture to lower the influence of noise, the contours of the objects seen in the picture are computed. Assuming that no object lies on top of another, it is possible to use these contours to divide the gray–scale picture into several sub–pictures, each containing only one object $\mathcal{O}$. All pixels that do not belong to one of these objects $\mathcal{O}$ are given the value 0.
- Detection of location:
  To evaluate the two–dimensional characteristic of one of these objects $\mathcal{O}$ the center of mass is calculated. This point can be used to describe the transposition of the object in the x–y plane referring to the coordinates of the sensor system.
- Process of intersection:
  Now, a rotational symmetric test object T is placed in the scene. The axis of rotation of this test object must be perpendicular to the x–y plane and the center of mass of $\mathcal{O}$ is located on the "special point" of the axis of rotation.

This rotational symmetric test object T must be generated in the training phase using the process described in 2.5. In the task of recognition, the test object T will be the one which was generated during the training.

The discrete intersection curve $\mathcal{I}(\gamma, h)$, gained by $\mathcal{T} \cap \mathcal{O} = \mathcal{I}(\gamma, h)$, is characterized by two parameters. The angle γ, used to describe the position of the intersection point in relation to the base circle of $\mathcal{T}$, and h, expressing the distance of the intersection point from the x–y plane. Using this parameter representation, the intersection curve $\mathcal{I}(\gamma, h)$ can be interpreted as a two–dimensional characteristic that fulfills the requirements mentioned in 2.2.

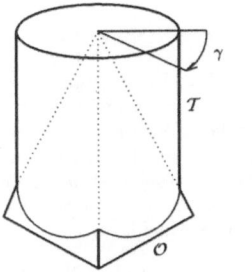

Figure 1: Intersection of the object $\mathcal{O}$ with the test object $\mathcal{T}$

2.3.2 Describing an object using the two–dimensional characteristic

The two–dimensional characteristic $\mathcal{I}(\gamma, h)$ does not change its general shape if the object $\mathcal{O}$ is transposed in the x–y plane or rotated around the z axis. If $\mathcal{O}$ is turned around another axis, the two–dimensional characteristic changes in a complex way. Therefore, this two–dimensional characteristic is only valid for one stable position of the object $\mathcal{O}$. To obtain a total representation of an object, the intersection curves $\mathcal{I}(\gamma, h)$ of all stable positions of the object must be computed and stored.

To evaluate the two–dimensional characteristics of a four–sided symmetrical pyramid, one has to compute two two–dimensional characteristics, one for characterising the pyramid standing on the ground plane and another one is needed to describe the pyramid lying on one side plane.

2.3.3 Algorithms used

Several well–developed algorithms which can be used for these three steps have been published. For the picture pre–processing described in step 1 we use median–filter and average–filter algorithms published by (Haberäcker, 1991) and (Rosenfeld and Kak, 1976). The evaluation of the contours of objects is done by a simplified version of an algorithm published by (Gagalowicz, 1985).

Computing the centers of mass of the objects can be done by using algorithms found in (Bronstein, 1981) and the evaluation of the intersection curve by using simple subdivision techniques.

2.4 Detecting an object

To test whether an object $\mathcal{O}$ is or is not the object $\mathcal{O}_i$ which was learned, the process of recognition must do the following steps:
- First, it must query the recognition database about the test object $\mathcal{T}_i$, stored during training, and the intersection curve $\mathcal{I}_i(\gamma, h)$ which had been evaluated.
- Using the same algorithms as in 2.3.1 1–3, the gray–scale picture is pre–processed.
- To test whether the object $\mathcal{O}$ is the object $\mathcal{O}_i$ which was learned, we have to check if the intersection curve $\mathcal{I}(\gamma, h)$ of $\mathcal{O}$ and $\mathcal{T}_i$ is the same as $\mathcal{I}_i(\gamma, h)$.

During this test it must be observed that the object might have turned around the z-axis. Such a rotation leads to a horizontal shift of the intersection curve.

To detect such a shift, it is necessary to compute the standard deviations of all possible shifted $\mathcal{I}(\gamma, h)$ and $\mathcal{I}_i(\gamma, h)$. This is only possible because the $\mathcal{I}$ functions are discrete. If any of these standard deviations is 0, $\mathcal{I}_i(\gamma, h)$ and $\mathcal{I}(\gamma, h)$ are identical and the shift vector describes the rotation angle of the object $\mathcal{O}$.

136

Computing all these standard deviations with a normal mono–processor computer takes a lot of time. But it is possible to speed up this process by using a multi–processor system — like a transputer system — and evaluate the deviations in parallel.

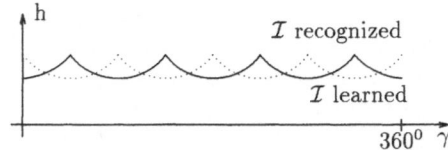

Figure 2: Shift of the intersection curve caused by a rotation of the object around the z axis

Whenever real sensor systems are used to recognize an object, it is impossible to work with numerically exact data. This is why the term "is 0" must be replaced by "is smaller than ϵ". The size of ϵ depends on the efficiency of the sensor system, the better it is, the smaller is ϵ.

2.5 Training the recognition system

Before an object recognition system is able to recognize an object, it must be trained. Using the two–dimensional characteristic described in the previous sections, training a new object $\mathcal{O}_{n+1}$ means choosing a suitable test object $\mathcal{T}_{n+1}$ and fixing the position of the "special point". A test object $\mathcal{T}_{n+1}$ is suitable if, and only if, the resulting intersection curve $\mathcal{I}_{n+1}(\gamma, h)$ is unique compared to all intersection curves $\mathcal{I}_{1,...,n}(\gamma, h)$ learned so far.

Before it is possible to compute the two–dimensional characteristic, the gray–scale picture of the object $\mathcal{O}_{n+1}$ has to be processed as described in 2.3.1 1–3. Once this has been done, the pre–processed gray–scale picture of $\mathcal{O}_{n+1}$ is stored in the recognition database, and a test object $\mathcal{T}_{n+1}$ for the computation of the intersection curve is chosen.

2.5.1 Requirements on $\mathcal{I}_{n+1}(\gamma, h)$

Before a new object can be trained and all information needed for the process of recognition stored in the recognition database $\mathcal{I}_{n+1}(\gamma, h)$, the following requirements must be fulfilled:

- The new intersection curve $\mathcal{I}_{n+1}(\gamma, h)$ must be unique:

$$\forall j = 1, \ldots n: \quad \begin{array}{rcl} \mathcal{O}_{n+1} \cap \mathcal{T}_{n+1} & \neq & \mathcal{I}_j(\gamma, h) \\ \mathcal{O}_j \cap \mathcal{T}_j & \neq & \mathcal{I}_{n+1}(\gamma, h) \end{array}$$

If this requirement fails, the test object $\mathcal{T}_{n+1}$ must be changed.
- The intersection of every object $\mathcal{O}_{1,...n}$ which was learned with the new test object $\mathcal{T}_{n+1}$ must not be the new intersection curve $\mathcal{I}_{n+1}(\gamma, h)$:

$$\forall j = 1, \ldots n : \mathcal{O}_j \cap \mathcal{T}_{n+1} \neq \mathcal{I}_{n+1}(\gamma, h)$$

If this happens for any object $\mathcal{O}_k, k = 1, \ldots n$, it is impossible to distingiush this object from $\mathcal{O}_{n+1}$ during the recognition process. To solve this, a new test object $\mathcal{T}_{n+1}$ must be created.
- The intersection of the new object $\mathcal{O}_{n+1}$ with any of the test objects $\mathcal{T}_j$, $j = 1, \ldots, n$, already used, must not be the corresponding intersection curve $\mathcal{I}_j(\gamma, h)$:

$$\forall j = 1, \ldots n : \mathcal{O}_{n+1} \cap \mathcal{T}_j \neq \mathcal{I}_j(\gamma, h)$$

If this happens for an object $\mathcal{O}_k$, it is again impossible to distinguish the new object and the object $\mathcal{O}_k$ during the process of recognition.
This problem arises because the information already learned (and therefore stored) leads to the inconsistency. This is why, simply changing the new test object $\mathcal{T}_{n+1}$ does not solve the problem. Two possible solutions are:
– The test object $\mathcal{T}_k$ could be changed. But in changing the test object of an object already learned, care should be taken not to destroy the consistency of the recognition database. In practice, this solution means extracting the object $\mathcal{O}_k$ and all related information, integrating the new object and than treating $\mathcal{O}_k$ as a new object. This means of guaranteeing the consistency of the recognition database is only useful if there is such a conflict for only one k.

– The second possibility is to create an additional test object to distinguish these two objects. Normally, this solution is chosen if the database is large or if the problem is not restricted to only one object already learned.

2.5.2 Generating the new test object $\mathcal{T}_{n+1}$

As a first step, an arbitrary test object for the intersection process is generated and then the intersection curve calculated. If this intersection curve fulfills all requirements mentioned in the previous section, the test object $\mathcal{O}_{n+1}$ and the intersection curve $\mathcal{I}_{n+1}(\gamma, h)$ are stored in the recognition database, and the object $\mathcal{O}_{n+1}$ is trained.

If the intersection curve fails to fulfill one of the requirements, a new test object must be generated. The new test object could be created from the old test object using an algorithmic approach. But it is not evident, how the algorithm should change the test object. This is the reason why we generate again an arbitrary test object.

Having selected the new test object, we must check whether the new intersection curve fulfills the requirments or not.

3 Conclusion

Using the method described in this paper, it is possible to recognize all kinds of objects by means of a gray–scale CCD camera. It is no longer necessary that the objects to be recognized should consist of simple geometric elements. This method fulfills the requirements mentioned in section 1.1. The demand of detecting objects in real time can be fulfilled by using multi–processor computers.

Furthermore, by calculating the intersection between the whole surface of the object and the test object. this approach also uses texture elements in the process of recognition. For this reason, the method can be extended easily to tasks of object recognition based on pictures obtained by color–ccd–cameras.

References

Besl P. J., Jain R. (1985). Three–Dimensional Object Recognition. Computing Surveys. Volume 17,/1. 75 – 145

Bronstein I. L., Semendjaev K. A. (1984). Taschenbuch der Mathematik. Verlag Harri Deutsch, Thun und Frankfurt(Main)

Gagalowicz A. (1985). A new approach for image segmentation. International conference on pattern recognition. 265 – 268

Haberäcker, P. (1991). Digitale Bildverarbeitung. Carl Hanser Verlag, München, Wien.

Lowe D. G. (1987). Three–Dimensional Object Recognition from single Two–Dimensional Images. Artificial Intelligence. Volume 31. 355 – 395

Perrott C. G., Hamey L. G. C. (1991). Object Recognition, A Survey of the Literature. Macquarie Computing Reports No. 91-0065

Rosenfeld A., Kak A. C. (1976). Digital picture processing. Academic press inc.. New York

Schrott, A. (1993). Ein Verfahren zur visuellen Unterstützung des Greifvorgangs bei Robotern basierend auf einer Greifer–Sensor–Koppelung. Technische Universität München. Doctorate thesis

Multi-sensor Technique for Increasing Intelligence of Assembly Robots

F. Alpek, Z. Nagy, P. Sallay, T. Szalay, K. Szelig, K. Toth
Department of Production Engineering
University of Technology, Budapest, Hungary

Abstract: The authors, who are experts of the Technical University of Budapest Department of Production Engineering, work on the solutions of monitoring and sensoring problems in the CIM Pilot System. The results and the further tasks are discussed in this paper.

Increasing the reliability and flexibility of the system were successfully realised with a shared intelligence monitoring system based on multi-sensor technique. It included monitoring and controlling the force in robot gripper, monitoring assembly operations with force/torque sensor integrated into the robot wrist, and visual supervising with image processing.

The result was to perform unmanned assembly processes with application of the different sensors mentioned above. Application examples representing these methods are also introduced in the final chapter of this article.

1. Introduction

For automation of assembly processes it is necessary to replace the manned inspection with sensors for realising an unmanned and poor supervising system, respectively.

Monitoring forces and torque's reacting during the assembly operation and visual supervising play an important role in assembly processes. Other solutions, like acoustic emission and vibration analysis, are mostly used for monitoring cutting operations. One monitoring way cannot be replaced with another, applying them together makes possible to supervise the whole assembly process reliably.

Using multi-sensor technique such numerous problems take place, which are unknown or unimportant by applying single sensors. The difficulties are the following among others: making decision algorithms and strategies, increasing the speed of signal processing, choosing priorities, developing appropriate software's, e.t.c.

2. Layout and tasks of the assembly cell

At the Department of Production Engineering of Technical University of Budapest research work has been done since 1987 in area of robotized assembly and quality control [1,8]. A CIM Pilot-System was set up by our laboratory in co-operation with the Institute of Mechanical Technology and Material Science between 1989-91. The layout of the CIM System is shown in Figure 1.

The assembly cell of the CIM Pilot System was set up based on the experiences of former developing projects. The basic task of the cell is the automation of complex assembly processes involving monitoring the whole assembly sequence and robot movements, controlling the quality of the assembled products, and outputting the results of the assembly. For the realisations of the assembly operations different mechanical units, e.g. drivers, grippers, tool changers, e.t.c. had to be developed.

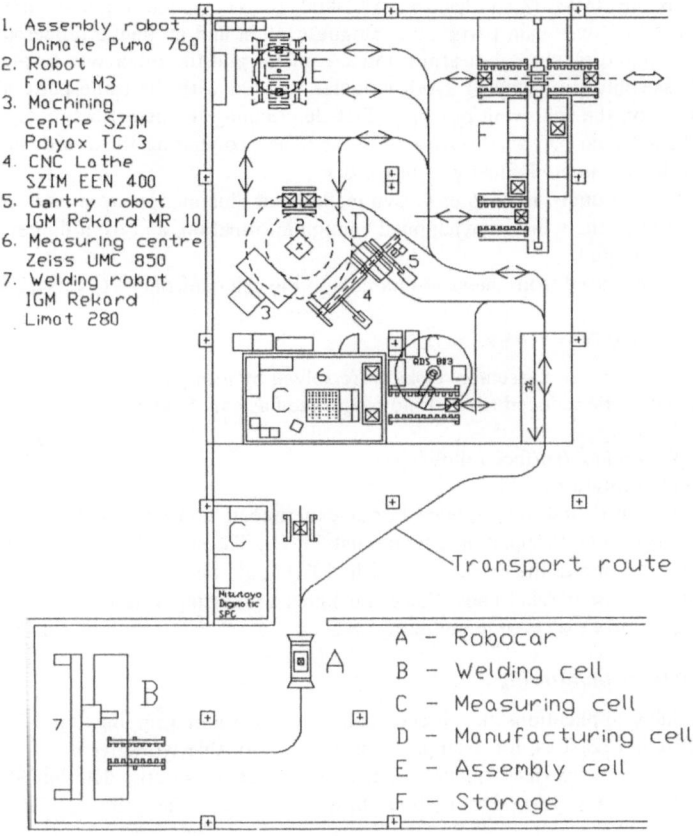

1. Assembly robot
   Unimate Puma 760
2. Robot
   Fanuc M3
3. Machining
   centre SZIM
   Polyax TC 3
4. CNC Lathe
   SZIM EEN 400
5. Gantry robot
   IGM Rekord MR 10
6. Measuring centre
   Zeiss UMC 850
7. Welding robot
   IGM Rekord
   Limat 280

Transport route

A - Robocar
B - Welding cell
C - Measuring cell
D - Manufacturing cell
E - Assembly cell
F - Storage

Figure 1. Layout of the CIM Pilot System of TU Budapest

For realisation of assembly automation a complete monitoring system based on multi-sensor technique was developed, which basically worked with force/torque monitoring and visual supervising based on shape recognition [2].

The three main monitoring tasks to be solved were the following:

- Force monitoring and control in the robot gripper [3]
- Force and torque monitoring with 6-axis force/torque sensor [5, 6]
- Image processing, shape recognition [4, 7]

3. Sensors used for increasing the reliability of robot

In our developments three different intelligent sensors were applied. Their structures and working methods are introduced shortly in the next chapters.

3.1. Vision systems

In the assembly cell the following main tasks could be solved with a vision system: identification of the workpiece, determination of the rough position/orientation of the workpiece and quality control of the selected part of the workpiece.

Two vision modules (VM-02 and Mikromat M-8001) were applied in the assembly cell. Each module consisted of two main parts: an analogue camera and an image processing unit. One camera was mounted on the robot arm (mobile camera) and the other was located above the place of the assembly (fix camera). Both modules worked with the same pattern recognition algorithm based on the following principle: first determining mathematical, geometrical, e.t.c. features of the selected part of the workpiece, and then comparing them with the features of the taught model by the predefined weight factors.

In the examples introduced in chapter 4. two methods of illumination were used. Back lighting when the contour of the workpiece supplied enough information, and front lighting when more information was needed.

The data was transmitted from the vision module to the cell controller PC via serial lines.

3.2. Six-axis force/torque sensor

In the assembly cell several assembly tasks were solved by using a six-axis force/torque sensor. Most of them could be reduced to the following three basic methods:
- peg-in-hole fitting
- accurate positioning (surface following)
- torque limited rotations

The sensory system (Miniforce) applied in our experiments consisted of two parts: a six-axis force/torque sensor and its signal processing unit (SPU). The sensor was mounted between the last wrist of the robot and the end-effector. The SPU could be connected to the cell controller PC via either serial or parallel lines. Since the signal processing was fast enough the robotic positioning tasks were executed in real time.

3.3. Gripping force monitoring

In some assembly applications the direct control of the robot gripping force is unavoidable because of different reasons, for example in case of deformable workpieces (like a thin-walled cylinder) or when the surface quality of the workpiece is strictly determined, e.t.c. Two different methods were carried out about how to measure the gripping force during the assembly.

One way is to place an external force measuring unit in the robot's environment. In this case the robot has to grasp this external load cell every time when it is needed and then continues the next assembly operation (pre-process measurement).

The other way is to make possible the continuous monitoring of the gripping force during the assembly (in-process measurement). In our experiments a PUMA 760 robot was used with a special gripper (having two fingers in symmetrical position) developed by us.

To measure the force in the robot finger a strain gage loading cell was integrated into one robot finger. The signal processing was solved by using a special PC card, so the measured force data could be directly read by the cell controller software.

Since the robot gripper is pneumatic the control of the gripping force could be realised by controlling the air pressure of the system. With the given hardware the air pressure could only be controlled by discrete steps (0.5 bar) instead of the continuous control, but it was appropriate from the application point of view.

4. Application examples

In this section four applications are described, which were realised by the authors at the Department of Production Engineering of Technical University of Budapest.

4.1. Assembly of IKARUS bus rear lamps

The task consisted of preparing the rear part of the bus for the assembly, pre-working the assembling place of the lamp-body and inserting the lamps into the appropriate position.

The first step of the robotic assembly was the determination of the position of the rear part. Pre-positioning (accuracy of 1 mm) and final-positioning (accuracy of 0.1 mm) was realised by applying vision module and f/t sensor, respectively. After it the holes of the four fixing screws were drilled by the robot having a drilling unit, finally the fixing screws of the lamp-body were screwed automatically, too.

4.2. Quality control of windscreen-wiper rotors during assembly

The major tasks were among others measuring the radial runout of the axis of the rotors and straightening the defectful workpieces. Measuring the radial runout of the rotors was performed by MITUTOYO DIGIMATIC incremental measuring instruments with μm precision. The straightening was carried out by a pneumatic gag press. The workpiece transportation was realised on palettes and roll-chain belt conveyor.

Some experiments were also made for controlling the position of the rotors based on visual information.

4.3. Robotized assembly of ball-taps

During assembling the ball-taps two vision-modules and a f/t sensor were applied partly for checking the workpieces in different steps of the assembly sequence, partly for supervising the technological operations.

The qualification (shape error, cutting error, e.t.c.) of the pieces to be assembled was solved by an image processing module. The visual information was also suitable for recovering the pieces from a position error, that could occur accidentally during the robotic operations.

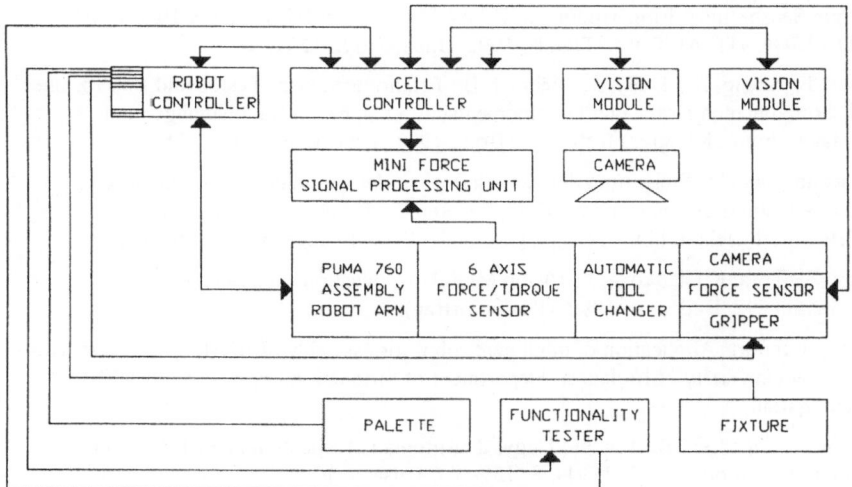

Figure 2. Information flow chart of the assembly of ball-taps

A six-axis force/torque sensor was used for monitoring the technological forces reacting during the assembly. With force control high precision positioning tasks, such as peg in hole fitting, were able to be solved several times, e.g. the insertion of assembled ball into its house.
The assembled taps were pneumatically tested for opening and closing. The information flow chart can be seen in Fig. 2.

5. Summary and further tasks

Using multi-sensor technique more information can be acquired from the assembly operations at one time, so the safety and the reliability of assembly can be increased, and the unmanned assembly can be performed.
The increase of the speed of processing the measured data and decision-making strategies require really hard software developments. Application of neural networks assures further advantages, that is one of the future tasks. Simulations of assembly processes including sensor functions are going to be worked out in the future.

6. References

[1] Alpek, F., Jerzsabek, P., Nagy, S., Nagy, Z., Szalay, T., Szebeni, J., Szélig, K., Zsembery, F.: Entwicklung einer robotisierten Meß- und Richtzelle. e&i, 108. Jg. (1991), Heft 6. S. 238-241.

[2] Alpek, F., Jerzsabek, P., Nagy, Z., Sallay, P., Szalay, T., Szélig, K.: Überwachungsaufgaben der Montagezelle im CIM-Pilotsystem der TU Budapest. e&i, 110. Jg. (1993), Heft 3. S. 127-134.

[3] Alpek, F., Péntek, L., Sallay, P.: Spannkraftüberwachung der Robotergreifer. 3. DAAAM-SYMPOSIUM "FLEXIBLE UND INTELLIGENTE AUTOMATION", 1992, Budapest, S. 5-6.

[4] Alpek, F., Nagy, Z., Tóth, K.: Increase of Intelligence of Assembly Robots with Force and Torque Sensor and Vision Module. 3. DAAAM-SYMPOSIUM "FLEXIBLE UND INTELLIGENTE AUTOMATION", 1992, Budapest, S. 125-126.

[5] Alpek, F., Szélig, K., Toldy, J., Weinper, B.: Erfahrungen beim Testen und Einsatz eines sechskomponenten Kraft- und Drehmomentsensors an einem Handhabungs- und Montageroboter. Kongress Robot '88, Brno, 1988. 1. Sektion, S. 171-175.

[6] Forschungsbericht: Steigerung der Zuverlässigkeit der Montageoprationen mittels Bildverarbeitungssystemen und 6-Komponenten-Kraft- und Drehmomentsensor. TU Budapest, Lehrstuhl für Fertigungstechnik 1990. (In ungarischer Sprache.)

[7] Nagy, Z., Tóth, K.,Gladkova, I.: Vision Module Applications in Assembly Cells. Automation '92, Budapest, 1992. (Postervortrag)

[8] Research report: Application of intelligent robot for assembly of IKARUS bus rear lamps. Technical University of Budapest, Department of Production Engineering, 1987. (in Hungarian)

[9] Szélig, K., Alpek, F., Berkes, O., Nagy, Z.:Automatic Inspection in a CIM System. Computers in Industry, 17 (1991), S. 159-167. (Elsevier)

[10] Alpek, F., Nagy, Z., Sallay, P., Szalay, T., Szélig, K., Tóth, K.: Multisensor technique for increasing intelligence of assembly robots. 2nd International Workshop on Robotics in Alpe-Adria Region, 1993, Krems (Austria), pp. Su.5.3-1.- Su.5.3-8.

Design of a Planetary Leg Mechanism Using Chebyshev's Optimization Method

E. Pennestri
Department of Mechanical Engineering
University of Rome "Tor Vergata", Rome, Italy

A. Di Benedetto, N. P. Belfiore
Department of Mechanics and Aeronautics
University of Rome "La Sapienza", Rome, Italy

Abstract. After a brief review of structural schemes of planetary leg mechanisms, an application of Chebyshev's optimization method to the kinematic design of a planetary leg mechanism is proposed. The graphs herein presented allow a quick proportioning of the mechanism for a given structural error or step lenght. The main advantage of this approach, with respect to other design methods, is the avoidance of noncircular gears for the minimization of structural error.

Introduction

Among the most recent fields of robotics, the one concerning the walking machines has been investigated for many reasons. In the medical field, for instance, walking machine can be adopted to provide mobility to patients with incapacitating disabilities. They can also be used for missions in hostile environments such as alien terrains, sea, radioactive regions, underground mines, and rural regions.

According to the rules of the *SAE Annual Walking Machine Decathlon (Cardenas-Garcia and Tsai, 1989)* a walking machine is defined as *a mobile, terrain adaptive system with several (8 or less) articulated mechanisms (arms and/or legs) which can perform defined tasks in static and dynamic environments.* It has been shown *(Bekker, 1960)* and *(Song et al., 1984)* that legged vehicles can be more energy efficient than wheeled or tracked ones, when travelling over rough and/or soft terrain. It is also evident that in such an environment the mobility of a walking machine can be superior to that of a wheeled one.

Geometry of legs is a crucial aspect of design since it affects the energy consumed per unit distance. In particular, it has been proved *(Song et al., 1984)* that an horizontal straight-line motion for the foot, with respect to the body, is a convenient requirement during walking, to minimize the energy consumed and load oscillations.

Many models of leg mechanisms have been proposed during the last decades *(Guanxiong Cha et al., 1989)* : pantograph mechanisms, Odetic's leg, four-bar type linkages, Peaucellier inversors, and planetary gear leg mechanisms, which seem to be among the most recent leg devices *(Guanxiong Cha et al., 1989)* and *(Shkolnik, 1989)*.

A previous work *(Guanxiong Cha et al., 1989)* provided two different solutions for planetary gear leg mechanisms. In the first one, an exact straight line for the foot motion can been achieved by using non-circular gear. However, such gears are expensive and difficult to manufacture. For this reason, a second approximate solution has been proposed. Non-circular gears can be replaced by circular ones whose sizes are chosen close to the average of the corresponding non-circular gears.

In this paper, an application of Chebyshev's optimization method to the design of a planetary leg mechanism is proposed. The general approach is described in *(Freudenstein, 1965)* and *(Di Benedetto and Pennestrì, 1993)* and relies upon the minimization of the structural error. In our application, the adopted methodology gives an high-precision straight-line for the motion of the foot. All the gears are supposed to be circular. Beside, it is possible to eliminate the middle gear by imposing a belt connecting the sun and the planetary gear.

Numerical results demonstrate that the approximation to a straight-line for the foot motion is higher than the one reached by using criteria described in *(Guanxiong Cha et al., 1989)*.

Planetary leg mechanisms

The optimization thechnique decribed in this paper has focused on a particular type of leg mechanisms. The single device is embedded into a more complex design in order to provide a real and safe walking to the body. For a stable walking, hexapode machines have been often choosen in all the cases examined by the Authors of this paper. The main advantage is that walking can be arranged in such a way that three non aligned foots are always in contact with the ground. The leg mechanism under analysis will be therefore regarded as one of the six legs which sustain the body, while the optimization of the control and coordinating system will not be taken into account.

A significant example of planetary leg mechanism is described in *(Tsai et al. 1989)*.. The assembly is composed of six gears; two gear carriers and a leg attached to the final gear. The gear ratios are chosen in such a way that the leg does only translate.

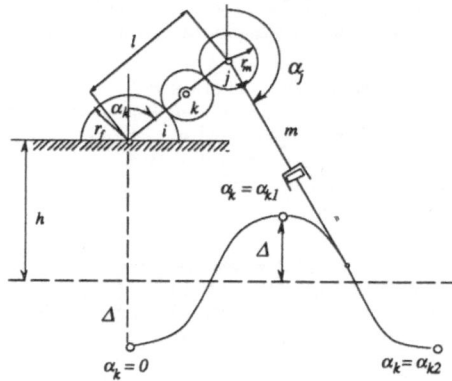

Fig.1: Scheme of the proposed planetary leg mechanism.

The proposed arrangement (see Fig.1) has a smaller number of moving parts and, if properly dimensioned, shows a negligible structural error (i.e. the tip of the foot describes a straight-line with an maximum error of 0.6% of the leg dimension, for all the cases considered) without the need of noncircular gears, as suggested in *(Guanxiong Cha et al., 1989)* and by *(Shkolnik, 1989)* in his U.S. patent application. The entire structure is rigid and compact. The area of the workspace can be varied by adjusting the length m of the leg during backward motion or obstacle climbing. However, the proposed layout requires the input gear carrier to be driven by a rocker. In our investigation this part of design has not been considered. For this reason m will be regarded as a constant.

Theoretical bases

With refer to Fig.2, the following notation is adopted:

- k, i, j :   subscripts which denote the gear carrier, body gear and leg wheel, respectively;
- Δ :   maximum value of the structural error;
- α_k :   angular position of input gear carrier link;
- $\alpha_j = \tau \, \alpha_k + \pi$:   angular position of the second gear carrier;
- ω :   angular velocity;

$-\tau = \dfrac{\omega_i}{\omega_k} = 1 - \dfrac{r_f}{r_m} :$ velocity ratio;

$- l :$ length of the gear carrier link;

$- m :$ length of the leg;

$- h :$ distance of the input shaft axis from the ground;

$- q = \dfrac{l}{m}$ normalized length of the input gear carrier link;

$- p :$ $\dfrac{\textit{Length of the half-step}}{m}$ normalized half-step length ;

$- L = \dfrac{\Delta}{m}$ normalized structural error,

Applying Chebyshev's conditions, one obtains the following design equations

$$q = \frac{1 - \cos \tau \alpha_{k2}}{1 - \cos \alpha_{k2}} \; , \qquad\qquad (6a)$$

$$\frac{\tau \sin \tau \, \alpha_{k1}}{\sin \alpha_{k1}} = \frac{1 - \cos \tau \, \alpha_{k2}}{1 - \cos \alpha_{k2}} \; , \qquad\qquad (6b)$$

$$L = \frac{1}{2} \left[\left(1 - \cos \tau \alpha_{k1} \right) - q \left(1 - \cos \alpha_{k1} \right) \right] \; . \qquad\qquad (6c)$$

Description of the algorithm and numerical results

By fixing the values of τ, α_{k2}, equation $(6a)$ and the numerical solution of $(6b)$ give, respectively, q and α_{k1}. Then, equation $(6c)$ allows the determination of the maximum structural error. The numerical results obtained for different choices of design parameters τ and α_{k2} are concisely given by graphs shown in Figures 3, 4, 5 and 6.

Such graphs allow a quick dimensioning of the system. In particular, one can adopt a maximum value of the structural error or the length of the step and then obtaining from the graphs the proportions of the entire mechanism.

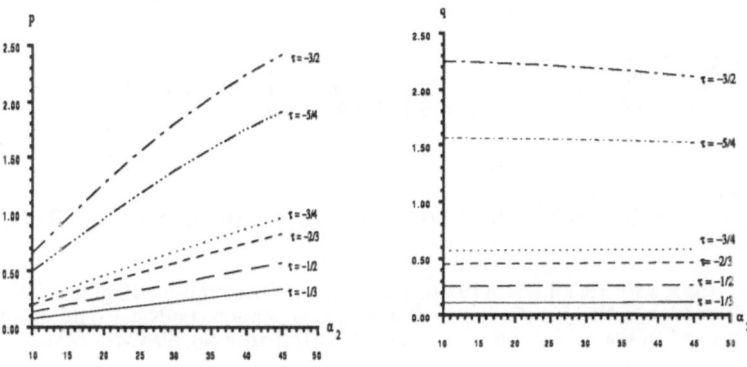

Fig.2 Half-step length p .vs. τ and α_{k2} .   Fig.3   Normalized length of input carrier q .vs. τ and α_{k2} .

146

In this last case, one should start from Figure 3, where the value of p is obtained by using the following equation

$$p = q \cos\alpha_{k2} + \cos(\tau \, \alpha_{k2} + \pi).\tag{7}$$

Otherwise, different criteria can be adopted, such as normalized input carrier length q or maximum normalized structural error L.

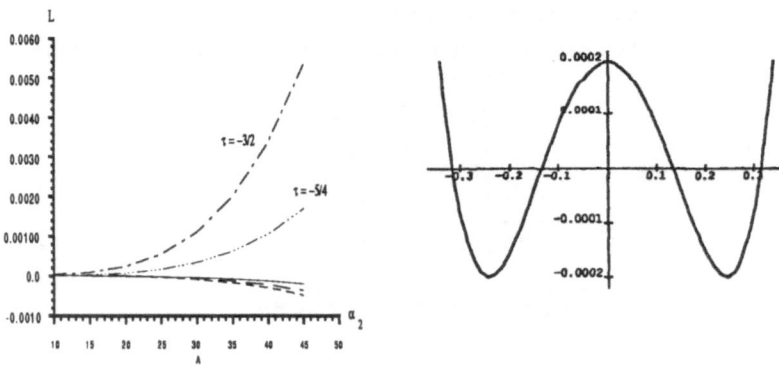

Fig.4 Normalized error L .vs. τ and α_{k2}.　　　Fig.5 Trajectory of the foot

Conclusions
An original approach to the dimensional synthesis of planetary leg mechanisms has been presented. The design equations are derived from Chebyshev's best approximation theory. The performed optimization takes into account both safety and energy efficiency of walking. In particular, a straight-line motion of the body with respect to the ground is granted with an approximation of 0.6%, relative to the length of the leg link.　In our algebraic treatment manufacturing errors are not take into account.
The comparison with other design methods confirm the feasibility of the proposed method.

Acknowledgement
The authors wish acknowledge financial support from Consiglio Nazionale delle Ricerche, Progetto Finalizzato Robotica, Contratto n.92.01089.PF67.

Bibliography
Bekker, M.G. (1960). Off-The-Road Locomotion, University of Michigan Press, Ann Arbor, Michigan.

Cardenas-Garcia, J.F., Tsai, L.-W. (1989). The Society of Automotive Engineers Walking Machine Decathlon, a National Collegiate Robotics Competition, 1st Nat. Conf. on Applied Mechanisms and Robotics, Cincinnati, Nov. 89, Vol. II, Paper No. 89AMR-7A-6.

Di Benedetto, A., Pennestrì E. (1993). Introduzione alla Cinematica dei Meccanismi, vol.1, Casa Editrice Ambrosiana, Milano, pp. 276-301, 390-395.

Freudenstein, F. (1965). Kinematic Synthesis of Rotary-to-Linear Recording Mechanism, Journal of Franklin Institute, vol.279, n.5, pp.325-333.

Guanxiong Cha, Ravi S. Saastry, Shin-Min Song (1989). A Comparative Study of Leg Mechanisms for Walking Machine Design, Proc. 1st National Applied Mechanisms and Robotics, Cincinnati, Ohio, vol.II, Paper No. 89AMR-7A-3.

Shkolnik, N. (1989).Walking Mechanism, United States Patent.

Song, S.M., Vohnout, V.J., Waldron, K.J., Kinzel, GG.L. (1984). Computer-Aided Design of a Leg for an Energy Efficient Walking Machine, Mech. and Mach. Theory, Vol. 19, No. 1, pp. 17 - 24.

Tsai, L.W., Chen, J., Azarm, S. (1989). The Design of a Three Degree-of-Freedom Walking Machine, 1st Nat. Conf. on Applied Mechanisms and Robotics, Cincinnati, Nov. 89, Vol. II, Paper No. 89AMR-7A-1.

A New Hybrid Locomotion Mobile Robot for Semi-structured Environments

C. Ferraresi, G. Quaglia
Department of Mechanics
Politecnico di Torino, Torino, Italy

Abstract: The article presents a small-scale prototype of a hybrid locomotion robot combining two types of mobility, i. e. on wheels and on legs. The prototype was developed from the authors' previous experience with a robot of the same type. Particular attention was devoted to simplifying kinematic layout in order to minimize the number of motors and provide the simplest possible control operations while ensuring a high degree of dexterity and speed. The article describes robot characteristics and the kinematic arrangements permitting motion on horizontal planes, ramps and stairs.

1. Introduction

Mobile robots are often used for operations in environments which were originally structured for humans. Examples of such operations include transporting objects, surveillance, monitoring, or emergency work in hazardous conditions.

Robot architecture must thus provide a full movement capacity in these environments. In particular, the robot must be extremely manoeuvrable on the horizontal plane so that it can avoid obstacles and go through doors and corridors. It must be able to negotiate ramps, if necessary maintaining its body at a constant trim, and must be capable of going up and down stairs.

In addition, the robot must have a sufficient operating range, which can only be achieved by minimizing the number of motors required for movement and by providing a compact, lightweight structure.

Moreover, if the robot is required to move in environments structured for humans, it is reasonable to assume that situations calling for locomotion on legs will be essentially restricted to going up and down stairs, whose geometry is almost constant. Rather than providing the robot with anthropomorphic legs, which are highly versatile and hence suitable for situations which are extremely demanding as regards locomotion, it is thus possible to equip the robot with a device designed specifically for stairs.

This article deals with a hybrid robot having the characteristics described above. The robot is equipped with a wheel unit for motion on level surfaces, whereby it can move along straight or curved paths and rotate on its own axis in an extremely small space. In addition, the robot is equipped with a device designed specifically for climbing stairs. This device consists of legs moved simultaneously by a single motor through an articulated parallelogram mechanism. The advantage of this solution is that the motor's continuous rotary motion can be used directly, so that the motor is only required to provide a relatively simple speed control. The legs can be readily adapted to different stair geometries, so that the robot is completely versatile.

Compared to previous hybrid robot prototypes (Belforte et al. 1990), where leg movement and geometry were much more complex, the solution presented herein made it possible to use only three motors for locomotion on both level surfaces and stairs. This provides obvious advantages in terms of light weight, reliability and controllability, in addition to increasing robot operating speed.

A small-scale prototype is now being constructed at the Department of Mechanics, Politecnico di Torino. This prototype offers a number of interesting features as regards performance, light weight and control simplicity.

2. Robot description

A view of the prototype design is shown in Figure 1.

Motion on level surfaces is achieved by means of two wheels W, which are driven by independent motors. The third support is provided by load sphere S. Using a sphere instead of a normal pivotting wheel ensures better trajectory control with a short radius of curvature while guaranteeing that support geometry remains constant at all times.

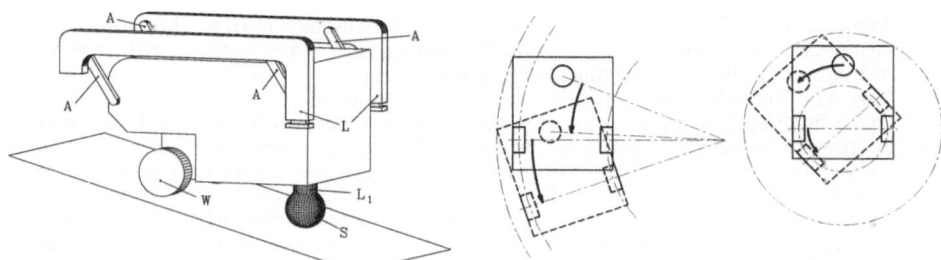

Fig. 1 - View of the mobile robot

Fig. 2 - Plan view of locomotion on
horizontal plane

Sphere S is located at the end of leg L_1, which can be extended vertically to permit movement along inclined planes with the robot body maintained at constant trim, or in order to adapt the structure for movement on stairs.

Movement up and down stairs also involves legs L, which can be extended vertically at the rear to adapt to step height. Legs L are moved by motor-driven arms A, which are connected to the robot body.

The robot is provided with a total of six DC motors, of which only three are used for locomotion. The remaining three automatically adapt the structure's geometry to the environment in which the robot must move. Of the three locomotion motors, two control wheel motion, while the third actuates the articulated parallelogram providing relative motion between legs and body.

The remaining three motors, which are of modest size inasmuch as they are active only during structure adaption, control vertical sphere extension and the two rear legs respectively.

The prototype was designed in small scale and is capable of negotiating stairs with a riser height of 60 to 80 mm and a length of 130 to 170 mm. Estimated total mass is around 30 kg.

A plan view of locomotion on level surfaces is shown in Figure 2. As can be seen, the two wheels are driven by independent motors, so that the motion trajectory is determined by controlling the speeds of the two motors. It is thus possible to produce trajectories whose center of curvature is located at any point of the straight line passing through the wheel axes. It is also possible for the robot to rotate around the vertical axis midway between the two wheels and thus change direction in the smallest possible space.

The motion of the robot along an inclined plane can be viewed again in figure 1. It will be noted that extending the leg carrying sphere S makes it possible to maintain the robot body at a constant trim.

The sequence performed by the robot in negotiating stairs is illustrated in Figure 3. This sequence consists of the following stages:

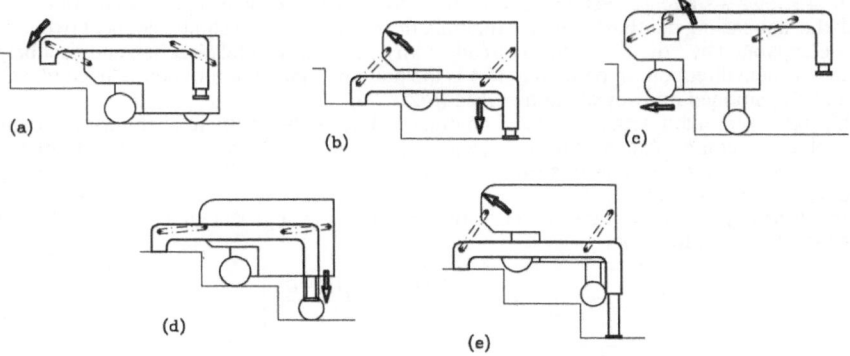

Fig. 3 - Sequence of movements in climbing stairs

a. Approach stage. The robot moves to a predetermined distance in front of the first step. It then rotates the parallelogram arms until the front legs rest on the first step, and extends the rear legs vertically until they touch the ground.

b. First stair. The robot lifts its body by rotating the parallelogram arms until the wheels rest on the first step. The leg carrying the sphere is then extended until it touches the ground.

c. Intermediate stage. The robot moves its legs forward while actuating the wheels to bring the body to a predetermined distance from the second step, compensating for any errors which may arise if tread width is not ideal.

d. Completion of adaptation operations. The front legs are brought to rest on the second step, and the rear legs are then extended to touch the ground. At this point the robot has adapted its structure to step height. If the latter remains constant, the motors controlling vertical extension of the three adaptable legs will not be actuated again until the end of the stairs.

e. Generic climbing stage. The motor controlling the articulated parallelogram now operates continuously. When the legs are resting on the tread, the body is raised and brought to the next step; subsequently, when the body is supported on the wheels and sphere, the legs are raised and the wheels are simultaneously actuated to position the robot relative to the next step, as described for stage c.

3. Kinematic analysis of stair-climbing

Once the kinematism involved in climbing stairs has been defined, a thorough kinematic analysis is required for mechanism sizing and design. Together, each leg and the two motor-driven arms form an articulated parallelogram. The two parallelograms are thus capable of generating translational relative movement between the robot body and legs with circular arc trajectories. This movement is divided into two stages: in the first stage, the legs are at rest and the body is moving, while in the second stage the body is at rest and the legs are moving. Both stages must be performed to negotiate a step. The following analysis focuses on the stage with the body in motion, as it is the most demanding in both dynamic and kinematic terms.

The arc radius and the angle at which body motion begins and ends depend on arm length and the position of the joints on the robot body. The characteristics of the mechanism will thus depend on the selected trajectory.

The trajectory must satisfy three basic requirements. First, it must ensure that there is no fouling between the robot and the step at any point. Second, it must make it possible to negotiate steps whose dimensions lie within a certain range. Third, the trajectory should not be too much in excess of riser height in order to limit the device's energy consumption.

These considerations were borne in mind in analyzing and designing a mechanism for the smallest staircase in the range selected for robot operation.

The smallest staircase was chosen since it can be seen it represents the most severe condition as regards interference between robot trajectory and staircase.

The effects of changing staircase dimensions on the robot's trajectory were then assessed.

Figure 4 shows a staircase section with the trajectory's basic geometric elements. In order to climb the stairs using continuous motor rotation, the robot trajectory should connect two points A and C separated by a distance equal to tread width p in the horizontal direction and riser height a in the vertical direction. In particular, if A is the point of contact of the robot's front wheel, it should be positioned midway along the tread $(p/2)$.

As the arcs represented with dashed and continuous lines indicate, there is an infinite number of possible trajectories. Amongst these trajectories, the one passing through point B at a distance d from the step was chosen, as it is sufficiently close to the step while still providing a margin of safety against collision.

The resulting expression for radius of curvature b and for the angles at which motion starts (θ_I) and ends (θ_F) are as follows:

$$b = \left(\frac{3}{4}p + \frac{d}{2}\right)^2 + \left[\frac{(p/2)^2 - pd - a^2}{2a}\right]^2 \tag{1}$$

$$\theta_I = \frac{\pi}{2} - \tan^{-1}\left(\frac{a}{p}\right) - \sin^{-1}\left(\frac{\sqrt{a^2 + p^2}}{2b}\right) \tag{2} \qquad\qquad \theta_F = \sin^{-1}\left(\sin\theta_I + \frac{a}{b}\right) \tag{3}$$

Radius b corresponds to the length of the parallelogram arms, while angle θ_I corresponds to the inclination which the arms must assume at the moment the front leg is placed on the step. This data makes it possible to design the arms and establish their position on the robot body.

Magnitudes b, θ_I and θ_F depend only on the dimensions used for the staircase and on the distance d between trajectory and staircase.

The following were adopted for the case at hand:

$a = 60$ mm, $p = 130$ mm, $d = 20$ mm.

These give:

$b = 90$ mm, $\theta_I = 12.5°$, $\theta_F = 118°$

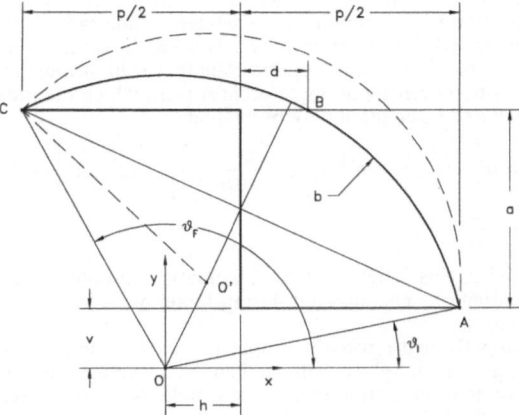

Fig. 4 - Basic geometric elements in stair climbing

We then analyzed what happens when stair dimensions are varied. As shown in Figure 5, for a riser height a' and tread width p', the end point of the trajectory should be shifted to C'', i.e. once again to the midpoint of tread, as no further wheel position correction is required in such a case.

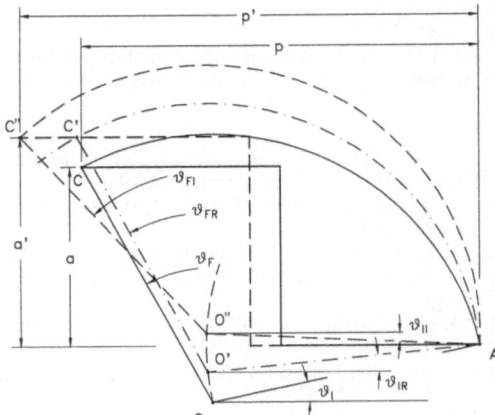

Fig. 5 - Kinematic behaviour with stair dimension variation

For a given arm length b, this would require two ideal starting and ending angles θ_{II} and θ_{FI} which differ from those given previously:

$$\theta_{II} = \frac{\pi}{2} - \tan^{-1}\left(\frac{a'}{p'}\right) - \sin^{-1}\left(\frac{\sqrt{a'^2 + p'^2}}{2b}\right) \quad (4)$$

$$\theta_{FI} = \sin^{-1}\left(\sin\theta_I + \frac{a'}{b}\right) \quad (5)$$

However, this can only be achieved by adapting the length of the front and rear portions of the legs.

In reality, as the front portion of the leg cannot be extended, the trajectory's new center of rotation is O', which is located at a vertical distance equal to $a'-a$ from O.
This gives two angles θ_{IR} and θ_{FI} as follows:

$$\theta_{IR} = \sin^{-1}\left(\sin\theta_I - \frac{a'-a}{b}\right) \qquad (6) \qquad\qquad \theta_{FR} = \sin^{-1}\left(\sin\theta_I + \frac{a}{b}\right) \qquad (7)$$

Several observations can be made at this point: angle θ_{FR} is independent of the actual stair dimensions a' and p', and is a function only of design geometry (a, b, θ_I). In fact, it corresponds to the configuration in which leg and front wheels are at the same level. As the robot body ends its trajectory at C' before starting to negotiate another step, it must move on wheels to C'' in order to reach a position corresponding to point A from which the previous circular trajectory began. The distance e to be covered may be evaluated as:

$$e = p' - b\left\{\cos\left[\sin^{-1}\left(\sin\theta_I - \frac{a'-a}{b}\right)\right] - \cos\left[\sin^{-1}\left(\sin\theta_I + \frac{a}{b}\right)\right]\right\} \qquad (8)$$

4. Conclusions

The robot described in this article features kinematic solutions which enable it to cope with almost all of the situations encountered during locomotion in semi-structured environments designed for human use.
Advantages deriving from the robot's simple movements and small number of degrees of freedom include high speed, light weight, and an extensive operating range.
In particular , movement on stairs is extremely simple, with clear advantages as regards robot speed, number and weight of actuators and control simplicity.
The considerations presented in the article provide the foundation for designing the stair-climbing mechanism, which is the most complex part.
A small-scale model is now being constructed at the Department of Mechanics of the Politecnico di Torino; estimated characteristics of this model are as follows:
- Total mass: 30 kg
- Overall dimensions: 0.3 m, 0.35 m, 0.25 m (width, length, height)
- Speed on level surfaces: 2 m/s
- Stair-climbing speed: 1 step/s
- Installed electrical power: 200 W
The prototype will be able to negotiate stairs of approximately half the normal size, and will be equipped with suitable sensors and control systems for experimental testing.

Bibliography

[1]   Ghee R. B. Mc. Iswadi G. I. (1979). Adaptative locomotion of a multilegged robot over rough terrain. In: IEEE Transaction on System, vol. SMC-9, n° 4
[2]   Okhotsimiski D. E., Platonov A. K. (1979). Integrated walking robot simulation and modelling. In: Proc. 7th IFAC, vol. 2.
[3]   Railbert M. H., Sutherland I. E. (1983). Macchine che camminano. In: Le Scienze n°175.
[4]   Takano M., Odawara G. (1983). Development of new type of mobile robot TO-ROVER. In: Proc. 13th ISIR, Chicago.
[5]   Ichikawa Y., Ozaki M., Suzuki M. (1983). Teleoperated mobile robot for remote maintenance in nuclear facilities. In: Proc. of 83 ICAR, Tokyo.
[6]   Saito M., Tanaka N., Arai K., Banno K. (1985). The development of a mobile robot for concrete slab finishing. In: Proc. 15th ISIR Tokyo.
[7]   Tsu-Tian Lee, Ching-Long Shih (1986). Real time computer control of a quadruped walking robot. In: Transaction of ASME JDSMC, vol 108.
[8]   Belforte G., Sorli M. (1987). Locomozione mista a zampe e ruote per ambienti semistrutturati. In: Proc. of AMMA Symposium on mobile robots, Torino.
[9]   Belforte G., D'Alfio N., Ferraresi C., Sorli M. (1988). Mobile robot with wheels and legs. In: Proc. Int. Symp. and expos. on robots. Sydney.
[10]   Belforte G., Ferraresi C., Quaglia G. (1990). Problems in locomotion of legged mobile robots. In: Proc. 21th ISIR, Copenhagen.

Commanding a Robot by Voice:
Speech and Autonomous Navigation for the
Mobile Robot of MAIA

B. Caprile, G. Lazzari

IRST - Istituto per la Ricerca Scientifica e Tecnologica
Trento, Italy

Abstract: MAIA is the project aimed at the integration of the AI resources presently operating at IRST. The overall approach to the design of intelligent artificial systems that MAIA proposes is experimental not less than theoretical, and an experimental setup (the experimental platform) has consequently been defined in which a variety of mutually interacting functionalities can be tested in a common framework. Here, the overall architecture of the system is outlined, and one of the platform's components - the Mobile Robot - is presented. Potentialities arising from the combined use of speech and autonomous navigation are also considered and discussed.

1   Introduction

MAIA (Italian acronym for *Advanced Model of Artificial Intelligence*) is the most important AI project presently developed at IRST (Poggio and Stringa, 1992). MAIA's long-term goal is twofold: (1) the realization of artificial systems exhibiting intelligent behaviours – being therefore capable of acquiring higher and higher levels of experience through interaction with humans and the external world; (2) to develop methods and tools which may allow to investigate artificial and natural intelligence from a rigorous experimental standpoint.

In order to provide the whole project with a common framework, an experimental set-up – the *experimental platform* – has been defined which can encompass a large portion of the AI research carried out in the Institute. Schematically (see Fig. (1)), the experimental platform of MAIA consists of "brains", that is a network of units collecting and distributing various kinds of information, and "tentacles" – remote devices devoted to the interaction with the environment in which the system operates. Three are the tentacles which today's experimental platform is provided with: the *Concierge*, i.e. a station interacting with humans in natural language (through a keyboard), and able to answer questions about the structure of the Institute, its organization or the scientific production of its members. The second tentacle is the *Electronic Librarian*, a station able to accomplish some of the duties any librarian has to carry out: to recognize the user from his or her face (Brunelli and Poggio, 1991; Brunelli and Poggio 1992), and the books from the cover (Stringa, 1991). The third tentacle of the system is the *Mobile Robot*. Beside assisting the Concierge and the Electronic Librarian in their functions (for example accompanying visitors from the reception desk to the offices), the Mobile Robot could also be asked to carry documents, books or other objects from point to point inside the building. The robot

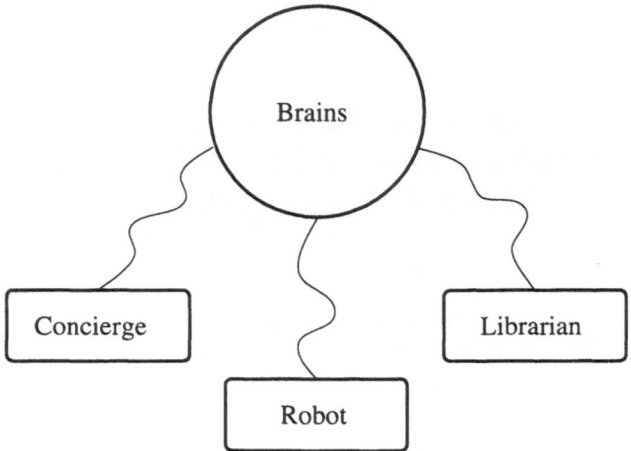

Figure 1: *The experimental platform of MAIA.*

can move autonomously in certain areas, and – beside a variety of other sensors and devices – it is provided with a loudspeaker and microphone to support vocal interaction with people (Cattoni et al., 1993; Antoniol et al., 1993).

Starting from year 1991 – when an early version was successfully tested for the first time – the system has undergone continual improvements. Although much work needs still to be done, speech and navigation modules cooperate today in an efficient and reliable way, under a wide variety of conditions.

2   The Mobile Robot

At present, the development of the robot takes place inside the building of our Institute, and the duties it is asked to accomplish consist in carrying objects from place to place or accompany visitors to the desired destinations. The system is therefore endowed with the capability of detecting particular structures of the environment, such as doors, elevators or corridors' crossings, and to perceive the presence of unexpected obstacles – people among them. The robot can also interact with people by voice. By voice an user can issue orders, give or ask for information concerning the missions, or finally help the robot, should it result unable to recover autonomously from difficult situations.

Two are the modalities (see Fig. (2)) in which the user can communicate with the robot: either in *standard* or in *telecontrol* modality. In standard modality, the operator imparts orders through a module – set aboard the robot – performing speech recognition and synthesis. Once that the acoustic signal is processed, results are sent to the Navigation Planner – a remote module living in the brains of MAIA. During this process, the user can be asked to solve ambiguities that the order may contain, or be asked for confirmation. The Navigation Planner finally translates the command

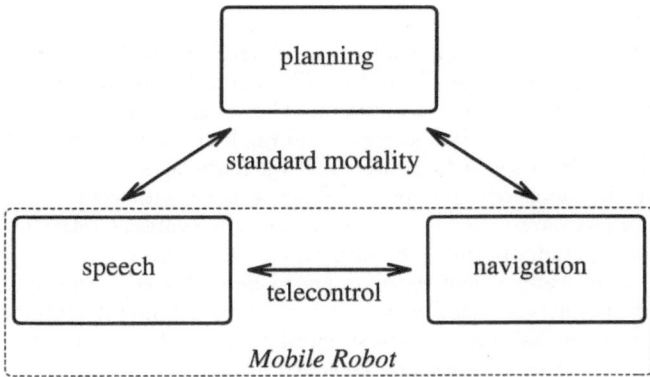

Figure 2: *Interaction modalities between the robot and the user. When the system is in* standard *modality, orders by voice that the robot receives come to it through a planning phase performed remotely. In* telecontrol *modality, a set of simpler orders can be issued bypassing the planning phase.*

into a *path* to be followed, and sends it to the robot. Aboard the robot, the path is translated in a series of elementary actions whose execution results, at least in principle, in the accomplishment of the mission. When switched to telecontrol modality, the robot is disconnected from the Navigation Planner, and, instead of a mission, it can be asked to perform simpler actions such as *"Move five meters along the corridor"*. In this case, no global information on the environment or abstract reasoning is needed, and the speech module interacts directly with the Navigation Module. Telecontrol can be particularly useful to operate the robot from a remote station, relying on the robot's own capability of coping with contingencies.

3 Speech Interaction

Research in the field of automated speech recognition has mostly been directed towards definition of sophisticated tools and techniques for dictation – that is, to obtain the correct sequence of words an user pronounces. Systems designed for this kinds of applications recognize without any reference whatsoever to the meaning of sentences. In this respect, they are not particularly appropriate to situations in which a context is well defined, and the semantic elements of sentences play a crucial role for recognition. Furthermore, to make the interaction with a robot friendly and flexible, *speaker independent* recognition of *continuous* speech is an almost mandatory feature for the system. Introduction of constraints on the admissible sequences of words may lead to a dramatic simplification of the problem. Considerations like these strongly support an approach to the comprehension of the spoken language based on models and decision schemes of probabilistic nature. However, this is no final solution, since other problems naturally arise; for example, that of creating a broad corpus of user-machine interactions from which values for the stochastic parameters of the model can be extracted.

A stochastic model of language allows to assign a probability to each sequence of words belonging to a certain vocabulary. Of course, it is not possible to estimate the probability of long sequences, since the number of examples required would rapidly exceed the volume of any database we could think of. Techniques for the estimate of the model of language have been proposed which are based on structural constraints represented by stochastic grammars. When dealing with comprehension systems, these techniques have to be used to model not only syntactic relationships, but also semantic and pragmatic ones (De Mori et al., 1991). Relationships among words – or sequences of words – and concepts are to be evaluated taking into account the context which sentences refer to. Let us take as an example the interaction between the Mobile Robot and the user in telecontrol modality. In phrases like *"move a little backwards"*; *"go around the obstacle keeping it on the right"*; *"stop here"* certain words or constructs will appear more often than others – that is, statistical parameters of models strongly depend on the context. Although it is only recently that research in this direction has been undertaken (Corazza et al., 1991), the results so far obtained are promising.

The recognition system developed at IRST is composed of the following modules: signal acquisition and acoustic parameters extraction; recognizer; interpreter; synthesizer. The signal, coming from a microphone is sampled at the frequency of 20 Khz and subdivided into "windows" of 20 ms. The beginning and the end of the phrase are then localized and the system computes the corresponding acoustic parameters. Starting from the acoustic description, a recognizer based on discrete Hidden Markov Models (Brugnara et al., 1992) determines the sentence belonging to the grammar which more likely correspond to the description. The interpreter, using the knowledge it has of the map and of the organization of the Institute, transforms the sentence it has recognized in a command for the robot.

The system has been trained on 25 speakers in conditions similar to those one can find in a typical office. The vocabulary consists of 200 words, and no pause is required to the user. An important problem we have started to tackle only recently is that of recognition of non linguistic phenomena or grammatically uncorrect sentences. Although some preliminary result has been obtained, much work has to be done before a user-machine interaction totally free of constraints is achieved.

Acknowledgements – The authors are deeply indebted to Renato De Mori and Tomaso Poggio for their crucial contribution to the research described in this paper. The authors wish also to thank Luigi Stringa for his constant support, and all the members of the Vision, Speech and Mechanized Reasoning Groups at IRST.

References

[1] A. Corazza and R. De Mori and R. Gretter and G. Satta. Computational probabilities for an island-driven parser. *IEEE Transactions on Pattern Analysis and Machine Intelligence*, 13:936–950, 1991.

[2] G. Antoniol, R. Cattoni, B. Cettolo, and M. Federico. Robust Speech Understanding for Robot Telecontrol. Technical Report 9302-19, Istituto per la Ricerca Scientifica e Tecnologica, February 1993. Accepted for oral presentation to ICAR '93 Conference, Japan.

[3] F. Brugnara, D. Falavigna, and M. Omologo. A HMM-Based System for Automatic Segmentation and Labeling of Speech. IRST Technical Report No. 9207-11, Istituto per la Ricerca Scientifica e Tecnologica, Trento, 1992. Published in the Proceedings of the International Conference on Spoken Language Processing [ICSLP 92]; Banff, Alberta, Canada, October 12–16, 1992; Vol. 1, pp. 803–806.

[4] R. Brunelli and T. Poggio. Face Recognition : Features versus Templates. IRST Technical Report No. 9110-04, Istituto per la Ricerca Scientifica e Tecnologica, Trento, 1991. To appear on *IEEE Transactions on Pattern Analysis and Machine Intelligence*.

[5] R. Brunelli and T. Poggio. Caricatural Effects in Automated Face Recognition. IRST Technical Report No. 9207-06, Istituto per la Ricerca Scientifica e Tecnologica, Trento, 1992. To appear on *Biological Cybernetics*.

[6] R. Cattoni, T. Coianiz, and B. Caprile. Planning, Reactivity and Learning for the Mobile Robot of MAIA. Technical Report 9303-06, Istituto per la Ricerca Scientifica e Tecnologica, March 1993. Accepted for oral presentation to ICAR '93 Conference, Japan.

[7] J. Baker. *Stochastic modeling for automatic speech understanding*. Morgan Kaufmann Publisher, 1990.

[8] F. Jelinek. Computation of the probability of initial substring generation by stochastic context free grammars. Internal Report 9111-09, Continuous Speech Recognition Group. IBM research, Y.J. Watson Center, Yorktown Heights, NY 10598, 1989.

[9] L.R. Bahl and F. Jelinek and R.L. Mercer. A maximum likelihood approach to continuous speech recognition. *IEEE Transactions on Pattern Analysis and Machine Intelligence*, (5):179–190, 1983. .

[10] T. Poggio and L. Stringa. A Project for an Intelligent System: Vision and Learning. *International Journal of Quantum Chemistry*, 42:727–739, 1992.

[11] R. De Mori and R. Kuhn and G. Lazzari. *A probabilistic approach to person-robot dialogue*. ICASSP, Toronto, Canada, May 1991.

[12] L. Stringa. Automatic book recognition. IRST Technical Report No. 9108-08, Istituto per la Ricerca Scientifica e Tecnologica, Trento, 1991.

Robot Welding of Tubes

P. Andris, K. Dobrovodsky, P. Kurdel
Institute of Control Theory and Robotics
Slovak Academy of Sciences
Bratislava, Slovakia

Abstract: A producer of tubes boilers initiated the solution of the following problem: A big diameter tube is to be welded with several tubes of small diameter. Tubes are to be welded by a robot along their intersection curves. To solve the problem, the authors modified the existing robot motion statements that were based on the interpolation of cylindrical coordinates. Trajectory generation algorithms for original and modified robot motion statements based on the quaternion representation of position are described.

1 Introduction

An integrated programming system for robot cells control has been designed at our institute. The system has been developed in cooperation with the Electric Engineering Research Institute in Nova Dubnica, Slovakia.

Sigma Slatina Brno, Czech producer of tube boilers, initiated the problem oriented modification concerned with the following problem: A tube is to be welded with several tubes of smaller diameter. Axes of all tubes lie in a plane. Axes of small diameter tubes are parallel. All axes of small diameter tubes are perpendicular to the axis of the big diameter tube. Tubes are to be welded along their intersection curve. Corresponding motion statements including an algorithm for trajectory generation are considered.

2 Quaternion Representation of Position

Our robot control system uses a quaternion representation of position (both translation and orientation), that has been proposed by (Dobrovodsky, 1985). (Andris, 1989) has used the representation for solving the inverse kinematics. The representation corresponds to the row vector convention of homogeneous transformation matrices. We will denote (m, n) the position of the cartesian system m with respect to the cartesian system n. We will represent the symbol (m, n) by the pair of quaternions $[r(m, n), R(m, n)]$, where

$$R(m, n) = \cos(\alpha/2) + iu_x \sin(\alpha/2) + ju_y \sin(\alpha/2) + ku_z \sin(\alpha/2)$$
$$r(m, n) = iv_x + jv_y + kv_z$$

The system n can be derived by the right-handed rotation of the system m around the fixed unit vector $u = (u_x, u_y, u_z)$ by the angle α. The rectangular cartesian coordinates of the origin of the system m with respect to the system n are (v_x, v_y, v_z). Let us notice that $R(m, n)$ is a unit and $r(m, n)$ is a pure quaternion. Under this notation, the following symbolic rules hold for positions of any three cartesian systems m, n, o:

$$(m, n)(n, o) = (m, o), \qquad (m, n)^{-1} = (n, m)$$

The multiplicative operation for the pairs of quaternions and the inverse of the pair of quaternions is defined by (1), where $\bar{T}$, $\bar{R}$ are the conjugates of quaternions of T, R, respectively:

$$[r, R][t, T] = [\bar{T}rT + t, RT], \qquad [r, R]^{-1} = [-Rr\bar{R}, \bar{R}] \tag{1}$$

3 Robot Motion Statements

The basic principle of the end-effector trajectory generation remains unchanged, i. e. motion statement is a statement of end-effector motion from a current position into the destination position. Destination position coordinates are a part of the motion statement and they are stored together with the line number and other parameters. The robot control system maintains three types of the current position coordinates. Joint coordinates represent the joint variables, cartesian coordinates represent the translation by a pure and the rotation by a unit quaternion with respect to the effector's base and with respect to the reference coordinate system. Cartesian coordinates with respect to the reference coordinate system are alternated with cylindrical ones. All three representations of the current position are kept current. One of the three representations is considered as "leading" and it is used for the trajectory generation. Other two are computed by using corresponding transformation equations. Coordinates of the motion statement destination position are stored with respect to the reference coordinate system, whose name is a statement parameter. These coordinates are either cartesian (if a joint or a cartesian interpolation is to be used) or cylindrical for a cylindrical interpolation. If the execution of the statement starts, it is not necessary to transform the destination position, but, if necessary, to transform the type of current position coordinates with respect to the reference coordinate system or to change the reference coordinate system. These transformations are executed (if the type of the trajectory is changed) during the starting phase of the statement execution. All motion statements are executed in the real time step by step. The number of steps is given either directly, or (if a motion velocity is a statement parameter) is computed on the base of distance to the destination position and the desired velocity. The further motion statement execution procedure is the same for all kinds of trajectories. Having executed one step, the new state is considered as an initial one, but the number of steps is decremented by 1.

The trajectory generation algorithm for joint coordinates, cartesian translation coordinates and cylindrical translation coordinates is very simple:

$$x_1 = \frac{x_n - x_0}{n} + x_0 \tag{2}$$

where is x_0 the initial coordinate

  x_n the destination coordinate

  n number of steps to the destination coordinate

  x_1 next step coordinate

If the velocity is a statement parameter, the number of steps to the destination position is computed again.

The orientation generation algorithm is a multiplicative form of (2) for unit quaternions

$$Q_1 = \sqrt[n]{Q_n Q_0^{-1}} \cdot Q_0$$

where is Q_0 the initial orientation

  Q_n the destination orientation

  n number of steps to the destination orientation

  Q_1 next step orientation

The algorithm of motion generation along a circle, a cylinder surface or a spiral is identical with the algorithm of motion generation along a straight line with a gradual orientation change. This is reached by generating in corresponding coordinates (cartesian or cylindrical) of a suitable located reference coordinate system. The only difference is that transformation of cylindrical coordinates into cartesian coordinates is executed during each step. This principle enables an easy implementation of generation of various types of trajectories by exchange a cylindrical transformation program module for another one. The cylindrical transformation equations are:

$$
\begin{aligned}
r &= \sqrt{x^2 + y^2} \\
\varphi &= \text{atan2}(y, x) \\
v &= z \\
W &= Q Z_\varphi
\end{aligned}
$$

where $Z_\varphi = \cos(\varphi/2) + k \sin(\varphi/2)$. Equations of the inverse transformation are:

$$
\begin{aligned}
x &= r \cos \varphi \\
y &= r \sin \varphi \\
z &= v \\
Q &= W \bar{Z}_\varphi
\end{aligned}
\tag{3}
$$

where x , y , z and r , φ , v are currently used cartesian and cylindrical coordinates and Q , W are unit orientation quaternions. The unit quaternion Z_φ represents a rotation by angle φ around axis z of coordinate system and $\bar{Z}_\varphi$ is a conjugate of the quaternion Z_φ .

4   Trajectory Generation for Welding of Tubes

From a mathematical point of view we are to generate a spatial curve that is an intersection of two cylinder surfaces. Let us assume that axes of small and big diameter tube are mutually perpendicular and lie in the same plane. The origin of the reference coordinate system lies on the tubes axes intersection and the y , z axis of the system lies on the big, small diameter tube axis, respectively (see Fig. 1).

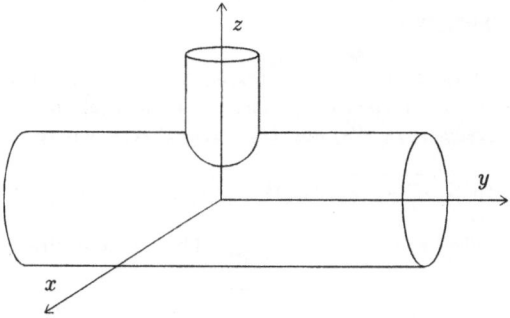

Fig. 1

If a cylindrical interpolation with respect to the system is used for the motion generation and the initial and destination point lie on the small diameter tube surface, a curve along the small diameter tube surface will be generated.

Let us suppose that the initial $(r_0, \varphi_0, v_0, W_0)$ and the destination position $(r_n, \varphi_n, v_n, W_n)$ lie also on the big diameter tube surface, whose equation is

$$x^2 + z^2 = R^2 \tag{4}$$

where R is the radius of the big diameter tube. Equations (3) and (4) yield

$$(r_0 \cos \varphi_0)^2 + v_0^2 = R^2, \qquad (r_n \cos \varphi_n)^2 + v_n^2 = R^2$$

Let us notice that coordinate r is constant during the interpolation, i. e. $r_0 = r_1 = r_n$. The orientation generation is not changed. The equation (2) is used to obtain φ_1 and v_1. The value of φ_1 remains unchanged, but v_1 will be corrected to satisfy the equation of the big diameter tube surface

$$(r_1 \cos \varphi_1)^2 + v_{c,1}^2 = R^2 \tag{5}$$

where $v_{c,1}$ is corrected v_1. The size and direction of correction $d_{v,1}$ is given by difference

$$d_{v,1} = v_{c,1} - v_1 \tag{6}$$

Equations (5) and (6) yield

$$d_{v,1} = sgn(v_1)\sqrt{R^2 - (r_1 \cos \varphi_1)^2} - v_1$$

where $sgn(v_1)$ is the sign of v_1. Because of inaccuracies during the teaching process, the initial and destination position of trajectory almost never lie exactly on the intersection of tubes. The inaccuracies can cause unexpected rapid movements of the end-effector and the generated trajectory can miss both the positions. To solve the problem, it is necessary to determine projections of the initial and destination position onto the big diameter tube surface. The projections are used to compute corresponding corrections $c_{v,i}$. Using of these corrections ensures the zero correction for the initial and destination position and eliminates unexpected rapid movements. The described modification has no effect if both positions are given exactly, but enables a user to make some inaccuracies during teaching of positions.

Let us suppose that $(r_0, \varphi_0, v_0, W_0)$ and $(r_n, \varphi_n, v_n, W_n)$ are really stored the initial and the destination positions, respectively. The projections of these positions onto the big diameter tube are $(r_0, \varphi_0, v_{p,0}, W_0)$ and $(r_n, \varphi_n, v_{p,n}, W_n)$, where

$$v_{p,0} = sgn(v_0)\sqrt{R^2 - (r_0 \cos \varphi_0)^2}, \qquad v_{p,n} = sgn(v_n)\sqrt{R^2 - (r_n \cos \varphi_n)^2}$$

The equation (2) is used to obtain r_1, φ_1, $v_{p,1}$. The size and direction of correction $c_{v,1}$ is given by

$$c_{v,1} = sgn(v_{p,1})\sqrt{R^2 - (r_1 \cos \varphi_1)^2} - v_{p,1}$$

The diameter of the big diameter tube must be inserted into the system. A non-zero value of the diameter signals that a modified cylindrical interpolation for welding of tubes should be used.

5   Conclusion

The paper deals with a quaternion representation of position and corresponding trajectory generation algorithms that are implemented in our programming system for robot cells control. A special robot motion statement that is designed for welding along intersections of two tubes is described. The diameters of the tubes may be different. The statement supposes that axes of both tubes are mutually perpendicular and intersect in a point. If these assumptions are not satisfied, only standard robot motion statements can be used.

Acknowledgement
The authors are grateful to the Slovak Grant agency for science (grant No. 2/999467/93) for partial supporting of this work.

References

Andris, P. (1989). Application of quaternions to inverse kinematics problem. In: Plander, I. (Ed.). Proc. of the 5-th International Conference on Artificial Intelligence and Information Control Systems of Robots. Elsevier Science Publishers (North-Holland), Strbske pleso, Czechoslovakia, pp. 359-363

Dobrovodsky, K. (1985). On the end-effector orientation generation. Computers and Artificial Intelligence, Vol. 5, No. 6, pp. 561-566

CAD for Robot Workcells in Battery Manufacturing

R. Kamnik, T. Bjad, A. Kralj
Faculty of Electrical and Computer Engineering Laboratory of Robotics
University of Ljubljana, Lubljana, Slovenia

Abstract: The article presents the initial step toward the introduction of robotics into the manufacturing process of the lead-acid batteries. Four possible applications of the AESA IRb 6 robot were studied using computer simulation support. The PC based robot simulation package WORKSPACE was used for designing and developing the robotic workcells. The package enables real time 3D graphic simulation of robots, conveyors, autonomous guided vehicels (AGVs) and other objects movements. The results of basic investment calculations for the four robot applications are also presented.

1. Introduction

Computer robot simulation seems likely to have an increasingly important role in the evolution of computer integrated manufacturing (CIM) facilities. The simulation packages provide a set of modelling and simulation tools which can be used to represent graphically a robot manipulator and its attendant equipment and hence simulate a manufacturing task. Some of the underlying reasons emphasizing the computer graphics simulation can be summarized as follows:

1. Anthropomorphic robot configurations tend to have many constraints, such as singularities and reduced path-velocity in the working space. Due to these constraints a real robot cannot perform every movement in the work space. A company cannot afford to buy several robot manipulators to compare their performance characteristics. The available simulation systems can provide a set of robot models which can be called up from their library.

2. Robot assembly cells are becoming more and more complex and may involve several robot manipulators, feeding devices, conveyors, fixtures and other machines. The selection of an optimal layout from many possible alternatives must be determined by relating the process and product information. The designer can easily make mistakes in ten-thousand dollar range when not carefully selecting the appropriate assembly devices.

3. The production process simulation can reduce to a large extent the time required for the implementation of a robot system: certain manufacturers claim that engineers can design and develop robot cells for up to 70% faster through the use of a robot simulator (Chan, 1988).

4. It is hard, tedious, and expensive task for robot design engineers to determine the structural, kinematic and dynamic characteristic properties of the robot manipulator without computer graphics simulation.

5. An additional feature of many robot simulators is the availability of postprocessing software for off-line programming of robots (Angermüller et.al., 1989; Parkin, 1991; Lee at.al., 1990). Such postprocessing is reformatting the geometric and sequential information generated by the modeler and simulator to produce a robot task program in the native language of the robot. In order to change products and test new programs, the automated production process has to be discontinued without using computer graphics simulation and off-line programming.

The robot simulation packages have proven in practice to be an effective and helpful engineering tool for design, analysis and control of robot systems (Mattis and Gill, 1988; Yocum, 1991; Janssen et.al., 1989).

In this contribution the robot computer simulation package WORKSPACE[R] is applied for support of modeling and simulating the four potential robot applications in battery manufacturing. The results of economic justification of particular application are also presented.

2. Simulation of a vibration machine serving

The vibration machine is used in battery manufacturing in order to fill the unfinished tubular positive battery plates (electrodes) with the lead oxide (Linden, 1984). Lead oxide is poured into the interior of the tube separator that is stuck to the plate lead grid. The empty plates are hanged up on a circular conveyor and carried through the vibration machine. When plates are filled, a plastic bung is put on the separator to close the orifice where the lead oxide has been poured into the plate. The finished positive plate is then taken from the conveyor and inserted into the output pallet.

The robot is aimed for serving the vibration machine. It is supposed to mount the plates on the conveyor at the right side of the machine, put on the bungs and take down the plates at the left side of the machine. The robot would replace two workers who are encumbered by enormous noise and presence of lead oxide. Both are breathing trough oxygen pipe for the whole working time. During carrying the finished plate to the pallet the plate weight with the robot wrist force sensor is also planned to be determined. Manual quality inspection of the plates would then be unnecessary.

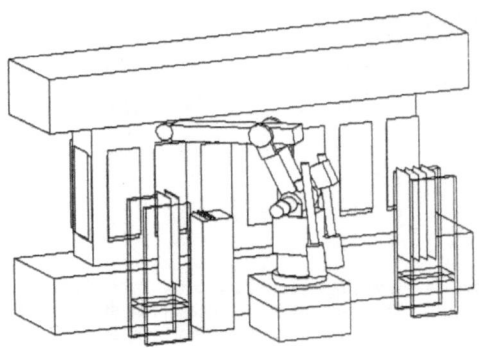

Manual quality inspection of the plates would then be unnecessary. The ASEA IRb 6 robot and vibration machine were modelled and defined in the WORKSPACE^R 'create-robot' environment. After modelling the robot, pallets and the bung stand the model of the robot workcell for filling the plates is designed. In figure 1 there is shown the ASEA IRb 6 robot during serving task at the moment when the bung is put on the plate.

For successful implementation of this task, it is necessary to overcome the problems like development of appropriate gripper for the plates and bungs grasping, the design of palettes and bung stand, the adaptation of the plate fixtures on the vibration machine for remote controlling.

3. Robot manipulation in plates formation task

Finished positive and negative battery plates are electrically formed in acid tanks. Plates are arranged in acid tanks in mixed succession. First, the negative plates are all inserted and then in between each of two negative plates one positive plate is placed. Both groups of plates are connected to a low-voltage, constant-current supply. Connecting is accomplished by placing and welding the lead bars on the plates plugs. After some hours, when the process of plates formation is finished, the lead connections are removed and plates are pulled out of the tank. The negative plates are put into the water tank and the positive plates are placed into the palette.

The acid tanks are arranged in 17 meters long rows, so the main requirement for successful robot plates manipulation is the mobility of the robot.

The robot mounted on AGV together with input and output pallets, connections stands and welding device is the proposal presented by simulation. Figure 2 shows the ASEA IRb 6 robot mounted on an AGV during filling the acid tank by the positive plates.

Protection of the acid influence on the robot mechanism and controller, selection of suitable AGV and strategy of AGV controlling, loading and unloading the pallets on AGV, welding device design and gripper design are the problems that are to be solved before completion of the task described.

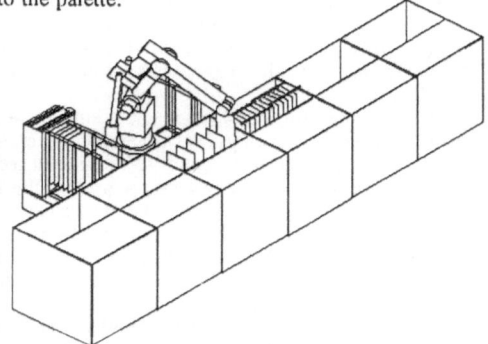

Figure 2: Plates insertion in acid tank

4. Robot manipulation during plates assembly

Before inserting into the casing, the plates are assembled in plate groups. The group of plates is composed from several positive and negative plates separated by separators and welded together. Each plate lug has to be brushed and clipped before welding can be accomplished. The plate lug brushing and lug clipping machine consists of a turning table, clipping press and brushing device. The plates on the turning table travel through clipping and brushing operations. This machine is rather difficult to use because of problems with appropriate loading and unloading of the turning table. The robot is aimed for carrying the plates from the input conveyor to the turning table and at the other side from the turning table to the output conveyor. The existent inadequate plates pushing pneumatic cylinders would be in this way replaced with robot loading system. Because of plate grasping instead of plate pushing, the machine would then be appropriate for both the negative flat-pasted and positive tubular plates. The output plate stack can be therefore composed in mixed succession what is convenient for further assembly. In the figure 3 the simulation of lug clipping and lug brushing machine with robot serving is presented. The plate has been just picked up from the turning table by the robot.

The design of convenient gripper for positive and negative plates grasping is the only difficulty of this application. Two modification are proposed: improved plate fixtures on the turning table and modification of brushing device with different brushes.

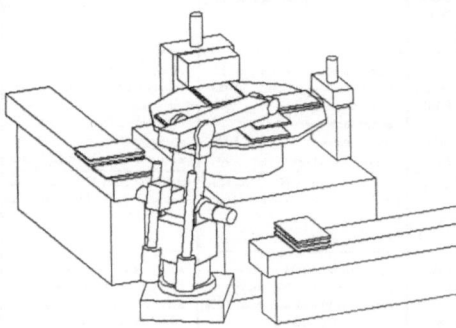

Figure 3: Plate picking up from turning table

5. Cells connection robot task

The final assembly of the traction batteries consists of connecting the finished battery cells that are placed in a plastic coated metal box. The battery cells are electrically connected by connecting links which are lead blocks with two chamfer holes on each side. The connecting links are put on two terminal posts of two different cells, thus the terminal posts are inserted into link holes. The connection is then fixed by welding. The positive and negative terminals are placed on the battery in the same way. In our simulation proposal robot puts on and welds the connection links. For overcoming the cell placement problems more complex sensory system should be applied, probably even a vision system. Figure 4 depicts the robot during connection link manipulation at the moment of putting connection link on the battery.

The expected projects for the implementation of this application are the design of gripper for connection links grasping, the design of a welding device with welding mould and the development of the connection links placement strategy.

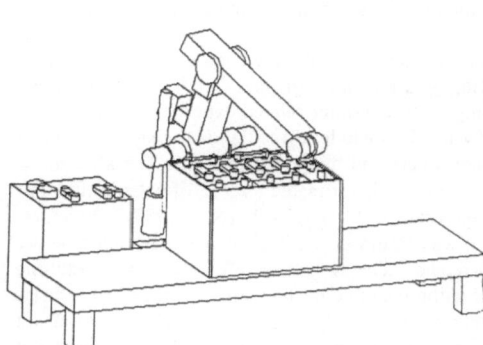

Figure 4: Cells connection robot task

6. Economic justification

Economic evaluation provides the decision framework to compare the benefits of automation through robotics with the present system and with other alternatives. The economic justification is based on the comparison between the capital cost and operating expenses of the robot installation that

	Vibration machine serving	Plates formation task	Plates manipulation during plates assembly	Cells connection task
Acquisition and start-up costs	- robot, controller 20000 - gripper 5000 - computer 5000 - force sensor 5000 - plate feeders - pallets 10000 - adaptation of vibration machine 10000 - installation costs 30000 - training costs 1000 - insurance 4000 total 89000 DM	- robot, controller 60000 - gripper 5000 - transport system (AGV) 70000 - welding system 10000 - equipment acid protection 10000 - computer 5000 - development of plates feeding system 10000 - installation costs 40000 - training costs 2000 - insurance 7000 total 219000 DM	- robot, controller 20000 - gripper 5000 - micro controller 2000 - adaptation of machine 5000 - development of plates feeding system 5000 - installation costs 2000 - training costs 1000 - insurance 4000 total 44000 DM	- robot, controller 20000 - gripper 5000 - sensors 15000 - feeding system 7000 - welding system 6000 - installation costs 4000 - training costs 1000 - insurance 4000 total 53000 DM
Operating expenses	- direct labor to tend robot 513 - supply pallets with plates 5146 - maintenance labor 209 - adjust tools for changeovers 3074 - energy and maintenance 6000 total for 1 shift 14942 DM/year total for 2 shifts 26284 DM/year	- direct labor to tend robot 513 - supply feeding system 30535 - maintenance labor 305 - adjust tools for changeovers 4611 - energy and maintenance 14400 total 50364 DM/year	- direct labor to tend robot 513 - supply feeding system 1053 - maintenance labor 209 - adjust tools for changeovers 3074 - energy and maintenance 6000 total for 1 shift 10849 DM/year total for 2 shifts 18098 DM/year	- direct labor to tend robot 513 - supply feeding system 1053 - maintenance labor 305 - adjust tools for changeovers 3074 - energy and maintenance 6000 total for 1 shift 10945 DM/year total for 2 shifts 17090DM/year
System benefits	- salary of two workers per shift divided by efficiency index (75%) and increased 10% because of vacations and 6% because of absence of health reasons 30740 - quality inspection 1537 total for 1 shift 31770 DM/year total for 2 shifts 63539 DM/year	- salary of nine workers divided by efficiency index (75%) and increased 10% because of vacations and 6% because of absence of health reasons 137409 total 137409 DM/year	- salary of one worker per shift divided by efficiency index (70%) and increased 10% because of vacations and 6% because of absence of health reasons 16144 total for 1 shift 16144 DM/year total for 2 shifts 32288 DM/year	- salary of one worker per shift divided by efficiency index (75%) and increased 10% because of vacations and 6% because of absence of health reasons 15267 total for 1 shift 15267 DM/year total for 2 shifts 30534DM/year

Table 1: Capital investment, operating expenses and benefits of four robot applications

is being considered and the financial benefits projected. The base of our justification method can be found in the literature (Nof, 1985; Maleki, 1991).

For each robot application the capital investment, the operating expenses and the system benefits are estimated and collected in the table 1.

The costs associated with robotic installation and start-up are mainly the costs of robot, controller, accessories (conveyors, pallets, feeders etc.), tooling (end effectors, grippers, fixtures), engineering costs, programming costs, installation costs, training costs and other related expenses (insurance etc.). The operating expenses associated with a robot typically include: direct labor to tend robot, labor to supply plates to feeders, adjusting tools for changeovers and maintenance labor. The assumption of the required time for a worker to spend tending the robot is estimated as 5 % of entire shift. For maintenance cost calculation it was supposed that the robot is 98 % reliable (Nof, 1985). Therefore, to determine the planned maintenance downtime, it was assumed that 2 % of the operation time per year will be used for preventive maintenance and repairs. The system benefits are mainly resulting from reduction in direct labor. The headcount reduction benefits are furthermore increased because of improved efficiency of the robot workcell and no losses caused by vacations and absence of health reasons. To determine which investment is most profitable we chose two calculations: capital recovery period and rate of return of investment. First estimate determines the possible term to pay back the invested capital. The calculation is performed by dividing the total investment amount and the difference of the total annual labor savings and the total annual expense for the robot including maintenance and tending labor costs. The second calculation, rate of return on investment, is suitable method of selecting appropriate alternatives by comparing profit ratio and the investment. For calculating the rate of return the same numbers can be used that were used for calculating the payback period. In table 2 the formulas are presented and the results of both payback period and rate of return.

Calculations are accomplished for two examples of utilization, either the robotic workcell is running during two shifts or during only one shift. This is highly dependent on market demands. Exception here is the plates formation task as the duration plates formation process is so long that there is no need for acid tanks loading for more than during one shift.

	Vibration machine serving	Plates formation task	Plates manipulation during plates assembly	Cells connection task
Payback period $\Upsilon = \dfrac{First\ Cost}{Average\ Profit\ /\ Year}$	1 shift: 5.3 years 2 shifts: 2.4 years	1 shift: 2.5 years	1 shift: 8.3 years 2 shifts: 3.1 years	1 shift: 12.3 years 2 shifts: 3.9 years
Rate of return $ROR = \dfrac{Average\ yearly\ profit}{First\ Cost}$	1 shift: 18.9 % 2 shifts: 41.9 %	1 shift: 39.7 %	1 shift: 12 % 2 shifts: 32 %	1 shift: 8.2 % 2 shifts: 25.6 %

Table 2: Payback periods and rates of return of investment for particular robot application

7. Conclusion

A preliminary study of four possible robot applications in lead-acid battery manufacturing was presented. Manufacturing tasks in battery production: vibration machine serving, plates formation task, plates assembly and cells connection task are chosen for robotization primarily because of human factor problems since workers are working in tedious and health hazardous working conditions. Each potential robotic workcell is considered like an island of automatization and not like a part of completely automatized production system. This means that before and after the robotic workcell no significant modifications are expected.

The idea of particular workcell operation and layout is presented by computer simulation support. Each workcell was modelled and animated in 3D graphics. The major problems encountered in implementation of a particular cell are outlined. Thus, before a robot is installed, the corresponding workcell is defined and the capital investment amount estimated. Investment estimation together with estimation of the operating expenses and robotization benefits was also accomplished. On the basis of this data the capital recovery period and the rate of return of investment were calculated for each robotized production task. This two methods represent only rough appraisals as there were not taken into account such factors as taxes, inflation, recession and market variations. As we can see from the table 2 the economic factors show that the plates formation task is the most appropriate for investment, however it requires the highest investment. Our proposal considering low start-up costs is the robotization of the battery plates assembly. The project is the least demanding and requires almost no production discontinuing during development. The final decision on the investment must be made after referring to available funds, resulting quality of production, and other management considerations.

References

Angermüller, G., Niedemayr, E. and Roth, N. (1989). Off-line programming and simulation of flexible assembly. Assembly Automation, Vol. 9, No. 2, 97-102

Chan, S.F., Weston, R.H. and Case, K. (1988). Robot simulation and off-line programming.
Computer-Aided Engineering Journal, Vol. 5, No. 4, 157-162

Janssen, P.J., Harrand, V.J., Wessling, F.C. and Choudry, A. (1989). Simulation of a flexible arm robot for space station applications. Simulation, Vol. 53, No. 1, 10-14

Lee, D.M.A. and ElMaraghy, W.H. (1990). ROBOSIM: a CAD-based off-line programming and analysis system for robotic manipulators. Computer-Aided Engineering Journal, Vol. 7, No. 5, 141-148

Linden, D. (1984). Handbook of batteries and fuel cells. Publishing McGraw-Hill Book Company, New York

Maleki, R. A. (1991). Flexible manufacturing systems - The technology and management. Publishing Prentice Hall, New Jersey

Mattis, P.A. and Gill, K.D. (1988). The best robot for the job: simulation can help decide. Industrial robot, Vol. 15, No. 1, 32-34

Nof, S. Y. (1985). Handbook of industrial robotics. Publishing John Wiley & Sons, New York

Parkin, R.E. (1991). An interactive robotic simulation package. Simulation, Vol. 56, No. 5, 337-345

White, R.B., Read, R.K., Koch, M.W. and Schilling, R.J. (1989). A graphics simulator for a robotic arm. IEEE Transactions on Education, Vol. 32, No. 4, 417-429

Yocum, M.G. (1991). Simulation slashes weld line design time. Industrial robot, Vol. 18, No. 4, 33-34

An Automatic Control on Band Saw Tool Vibrations in the Primar Cutting Process

Z. Trpovski[1], S. Loskovska[2], D. Mihajlov[2], S. Suselski[1]

[1]Forestry Faculty

[2] Faculty of Electrical Engineering

The "Cyril&Methodius" University, Skopje, Macedonia

Abstract: The most important problem in the forestry industry is to obtain better utilization of the wood material. In that context, the questions as which are parameters influencing the production and how they can be controlled are becoming significant. The log as the first material changes a lot in the production process, so the efforts are to obtain an optimal utilization of this material. We work on primar cutting process parameters investigation and their automatic control. This paper contains the results of the investigation for an automatic control on band saw tool vibrations in the primar cutting process.

1. Introduction

Quantity and quality of semi-final products obtained in the primar cutting process are very important for the whole wood production process. Quality primar cutting process lead to greater quantities of the semifinal products and better utilization of the raw material. The main purpose of our work is to obtain an optimal primar cutting process.

It is important notice that primar wood cutting is a process where waste of material is the greatest in quantity. Several kinds of products are obtained by the process, like planks, bars and shavings. Planks and bars appear as more significant for futher production process, and of this point of view shavings are waste material. The purpose of primar cutting process is to obtain greather quantity of these products which fullfil requirements for their quality, length, surfaces and so on.

A lot of parameters inflience quantity and quality of primar cutting process. Tool vibrations are one of the most important parameters in the process. They are directly dependent from wood cutting high, transporting velocity, active tool working time, and so on. Tool vibrations are blade deviations of its balance state and produce planks and bars tickness variation and larger quantities of shavings. Increased cutting resistance which appears between connected surfaces of wood and tool in this case make tool more dull, produce undesirable thermal effects and tool steel changings. The final results of the extreme tool vibrations besides poor cutting performance lead to catstrophic blade and tool failures.

The direction of our work was to measure all parameters which lead to tool vibrations and their automatic control. Transporting velocity, cutting high and tool vibrations were measured. The obtained values for the parametres were processed by PC. The results lead to a conclusion that the tool vibrations can be controlled mostly by transporting velocity.

This paper contains the results of our investigations. A solution for an automatic control of tool vibrations is shown. Conection between system elements and a software requirements are described. The problems which appear in the working place application are discused too.

2. Computer determination of the primar cutting process parameters

An extensive investigation have been performed to get more exact informations about blade tool vibrations during cutting process. In the primar wood cutting the most frequently used machines are log band saws which enable different cutting high without blade changing, lower kerf waste, higher cutting speed and so on. Band saws also give an economical return on the raw material, and with proper blade selection and good maintanence they will give a smooth surface and a tight dimension control. According to that Luis - Brenta 1400 log band saw was chosen for parameter measurement and process optimization.

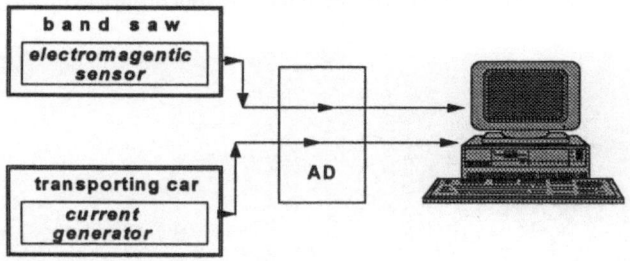

Fig. 1   A block diagram of the measurement system

Tool vibrations, transporting velocity, tool working time and cutting high were measured. The fig. 1 presents the connecting schema of the measurement system. An electromagnetic sensor is used to measure tool vibrations. Transporting velocity was measured by a current generator. These two instruments were connected on PC by A/D convertor. Cutting high was devided into classes. All investigations were performed on working place. Two types, pinus and fagus logs are used for investigations. These reasons were significant for their use in our work. They are:
- pinus and fagus woods are representative for our forestries; and
- the fist wood is an example for a soft wood and the second one is an example for firmly wood.

All logs were with similar characteristic acording to quality class, log diameter, humidity etc. Data were recorded for groups of 5 pinus and 5 fagus logs for active working time of 32 minutes with data sample rate of 50 /sec.

Analog signals obtained by measurement instruments are transfered by A/D convertor to PC and collected on hard-disk. The real values of the parameters are gotten by number of impulses collected by the computer.

All data were devided into several classes and statistically analizied. Analyses were performed for 5 classes for transporting velocity, 7 classes for cutting high and 4 classes for tool working time. The tables 1 and 2 show same of obtained results for pinus and fagus logs. Regresion analyses of results was performed and relation between tool vibration and the other parameters is obtanied. Fig. 2 and fig. 3 show relation between tool vibration and the other parameters for first and last four minutes. Results for pinus logs are presented on fig. 2 and fig. 3 represents results for fagus log.

Graphics show that vibration amplitudes increase with cutting high, transporting velocity and tool working time. The above conclusion lead to an idea to control vibration amplitudes with parameters changing. While a cutting high and a tool working time are still among the most desired characteristic, the transporting velocity is one which can be controlled.

3. Tool vibrations automatic control system

Tool vibrations automatic control system should insure getting of semi-final products dimensions with desired values. These means obtaining the same dimensions for whole length and surface of the planks and bars with allowing degree of correctness.

Maximal machine productivity and wood material and working energy saving should be insured too. The system should keep the tool of damage and insure tool blade sharpness control too.

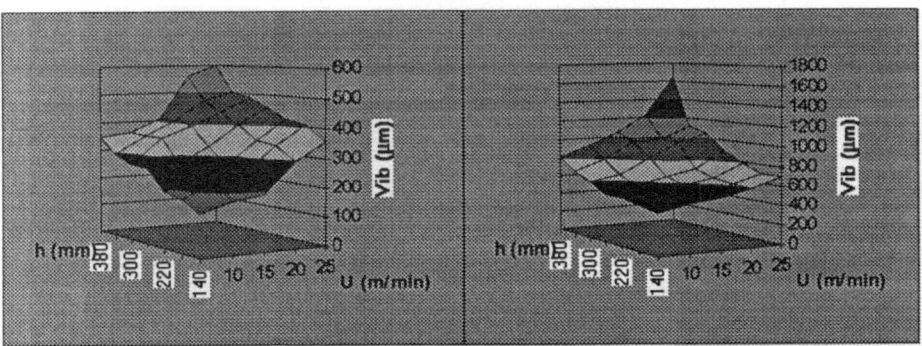

fig. 2 Relation between main values of tool vibration and the other
parameters for pinus log

Based on above requirements and conclusions described in previous pharagraph the following solution for system is obtained. Fig 4 presents the system elements connection.

A software for a given system solution is under development. It consists several moduls which insure:

- a process situation scanning;
- parameters calculation;
- values comparasion between momental and optimal process parameters;
- action choosing based on comparasion results

For above software moduls proper procedure work, these databases are designed:

- a component database;
- a knowledge database; and
- parameters database.

A component database contains all parameters information for a machine, measuremer instruments and environment. Parameters database consist tables with optimal values for eac tool vibration. The knowledge database contains relation between tool vibrations parameters and transporting velocity changing value.

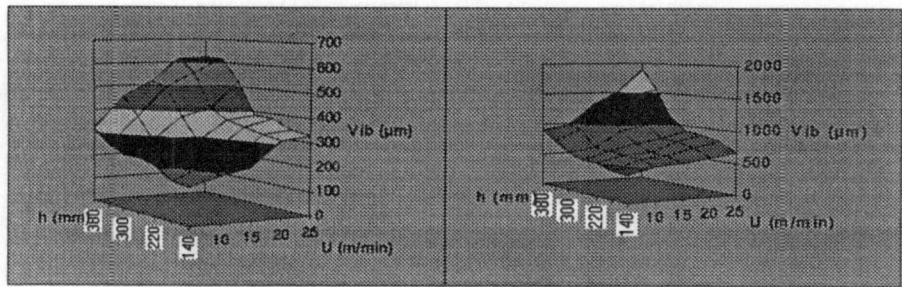

fig.3 Relation between main values of tool vibration and the other parameters for pinus log

4. Problems which appear with system application

There are several situations which lead to problems with system application. The main problem is that the optimal parameters are sensitive for a given machine and environment. In fact the optimal values are directly dependent from working place, wood quality, machine type and etc. So the offered solution give good results for this working environment and machine. For other machine, it is neccessary to perform new knowledge base. At this moment we are working on optimal tables valid for other machines.

Another problem which appears is data sampling rate. In this moment system works with data sample rate of 200/seconds. But the main problem is quantity of unuseful parameters values which computer get from its inputs. Only maximal tool deviations are significant for tool vibrations control, so a lot of computer time is lost on problem to eliminate unuseful values. The problem can be solved if there exists a component which will record only the useful values.

System inertness when knot is appearing is another problem. In that situation because of the mechanical log characteristic changings the tool vibrations increase and the system reacts with transporting velocity reducing. But this situation isn't stable for a long time and lower transporting velocity reduce machine productivity. These errors can be improved with action delay.

5. Conclusion

A solution for an automatic tool vibrations control is presented. This solition should improve primar cutting proccces and lead to better utilization of the raw wood material. This means that the process will insure obtaining greater quantity of semi-final products with better quality. As a big problem still remains systems sensitivity to the machine type and the working environment. Our reasearch is concentrated on obtaining a global system solution for differnt machines. The another problems like obtaining an optimal sample rate and making system to

172

be inertness are under development. The next step of our work is to applicate this system on different working place and imporove all possible problems.

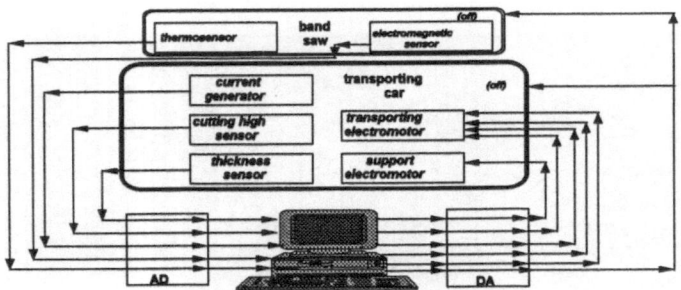

Fig. 4 An automatic tool vibrations control system for band saws

References

Breznjak, M., Mo en, K., (1972), "On the Lateral Movement on the Band saw blade Under Various Sawing Conditions", Norsk Treteknisk, Inst. Rep., No.46.

Broch, J.T., (1984), "Mechanical Vibration and Shock Measurements", Bruel and Kjaer Book Company.

Hribar, J., (1968), "Forces on tool for wood treatment", IV Conference for mechanica engineering, Sarajevo 1968.

Hutton, S.G., (1991), "The Dynamics of Circular Saw blades", Holz als Rohund Werkstoff 49 105-110.

Klincarov, R., (1978), "Zavisnosta na vibriranjeto na piloviot list i kvalitetot na rezenjet kaj lentovidna pila-trupcara od visinata na rezenjeto i debelinata na sticite", Sumarsk pregled 3-4, 43-58.

Pahlitzsch, G., Puttkammer, K., (1975), "Schnittversuche beim Bandsagen", Holz als Roh- un Werkstoff 33, 181-186.

Pahlitzsch, G., Puttkammer, K., (1976), "Schnittversuche beim Bandsagen", Holz als Roh- un Werkstoff 34, 17-21.

SANDVIK Steel publication, (1988), "Care and maintenance of wood band saw blades", Theor and Application Handbook, Sweden.

Ulsou, A.G., Mote, C.D., Szymani, R., (1978), "Principal Developments in Band Saw Vibratio and Stability Research", Holz als Roh- und Werkstoff 36, 273-280.

Low-cost CAD/CAM-coupling Applications
of Artificial Intelligence to Robotic Assembly

A.G.Yarmosh, A.P.Gavrish
Mechanics Machine-Building Department
Kiev Polytechnic Institute, Kiev, Ukrain
M.Zauner
Department of Systems Engineering and Automation
Scientific Academy of Lower Austria, Krems, Austria

Abstract: This paper deals with the problem of low-cost design-to-production integration. Due to the fact that there are now over 1/2-million AutoCAD users, including most of small companies, a new approach to CAD/CAM-coupling is proposed to connect an AutoCAD assembly drawing file as input with CNC production machines and assembly robots as output within one automated informational flow. Since interpretation of any real assembly drawing is strongly creative engineering task which until now has no satisfactory software solution, an Artificial Intelligence application in Prolog is involved. Case study illustrates automatic understanding of assembly drawings. Suggestions are made about fields for further applications, namely automatic determination of product's kinematics and robotic assembly sequences, generation of product's subassemblies and parts drawings, designer assembly production knowledge support.

1. Assembly Drawings Interpretation

At the earliest stage of design process designer makes first a detailed assembly drawing of the product. Only then he can make drawings of parts to enable their production. That is why the first problem for CAD/CAM integration in robotic assembly is implementation of automatic understanding of detailed assembly drawings. To "read" the drawing means not to enumerate and describe its graphical and text components, what is done by most of available CAD systems now, but to understand its contents good enough to solve the specified task: to clear out the components of the product, their relative arrangement, junctions, kinematics and other data necessary for planning and execution of assembly as well as production of parts. It is not easy but realistic task for every detailed drawing, since designer must execute each so exactly, that no additional comments become necessary (and if any appear, they must be found out and solved at the earliest steps of the design process with the help of such a system).

Automatic Interpretation of industrial assembly drawings as selection of separate components (parts) which are presented by their 2-D projections, and establishing kinematic relations among them, is creative intellectual problem in the meaning that it cannot be solved by algorithmic methods only. Although there are strict rules as to how to make and put assembly drawing into shape, assigning meaning to most graphical elements is dependent on context. Only methodology of Artificial Intelligence can give now tools to find out real technical objects with their actual form, orientation and location in space in such complicated for processing picture as assembly drawing. In this paper some results of the joint project of Kiev Polytechnic Institute and Scientific Academy of Lower Austria are discussed.

Since AutoCAD was selected as a source of input information, decision was made to create a stand-alone program able to estimate and use results of designs through analysis of AutoCAD output files. AutoCAD allows to skip video input and to go directly to reasoning about the 3-D image of a product in terms of its surfaces, volumes, solids and then parts. The Assembly Drawing Interpreter (ADI) program presented here, works with completed and ready for analysis drawings, which may be used, checked and even corrected and then returned to AutoCAD, if necessary. Under AutoCAD drawing we understand file with description of graphical primitives. Among formats of AutoCAD output files (DWG, DXF, IGS etc.) the most suitable for automatic processing and interface between AutoCAD and Prolog is DXF (Drawings eXchange Format), as it (1) contains all information available, (2) represents it in easy-to-read-by-PC way and (3) allows to generate/load drawing as a text file by simple "DXFOUT" and "DXFIN" commands by AutoCAD. Since each such file contains not only descriptions of graphical primitives of a product, but also complete information about the environment of the design process, abundant for this application, the first task solved was to ground selection of necessary entities and ensure their translation to the Prolog database. As a result a condensed mathematical representation substitutes raw input AutoCAD data, in the form of Prolog database facts:

entity(Entity#,InBlock#,Type={LINE|CIRCLE|ARC},View={FRONT|TOP|LEFT},LineType,
  {Xcen|Xmin},{Ycen|Ymin},{R|Xmax},{Side|Ymax},A,B,C,LIST)

where entities are described:
LINE - by *Xmin,Ymin,Xmax,Ymax,A,B,C,LIST=[Width,X0,Y0,X1,Y1]:*
  A,B,C-coefficients of the line equation $Ax+By+C=0$,
  Width={-1|0|1} if the line is *{thin|no_information|thick}*
  X0,Y0,X1,Y1-initial coordinates from DXF;
CIRCLE - by *Xcen,Ycen,Radius,LIST=[Width]*;
ARC - as part of *CIRCLE* cut by *LINE* on the indicated *Side*, according to $(Ax+By+C)*Side>0$, using
  Xcen,Ycen,Radius,Side={-1|0|1},A,B,C,LIST=[Width,A0deg,A1deg,X0,Y0,X1,Y1,Xm,Ym];
  A0deg,A1deg-initial starting and ending angles from DXF,
  X0,Y0,X1,Y1,Xmid,Ymid-coordinates for starting, ending and medial points on the *ARC*.

The first module of ADI transforms DXF files to Prolog internal database and Prolog database file. Additional feature allows to execute conversions from DXF to easy-to-read description file and back: this way any changes to a drawing can be made by designer out of AutoCAD.

The interpretation process must reveal all parts that are present at a drawing, i.e. tell the number of parts and their 3-D geometric images. To reach this goal, 3-D informational model of the product must be derived from assembly drawing. The model proposed consists of the following descriptions pyramide: entities-surfaces-volumes-solids- parts-product. First from the analysis of lines suggestions are made as to what surfaces does these lines represent (each line must be assigned to at least one surface). At the next step attempt is made to decide what surfaces compose closed simple volumes (for example, cylindric surface and two planes may compose a cylinder volume; some surfaces may belong to more than one volume). When these volumes are found out, decision is made on each whether it must be included to some solid body or excluded from it, or both - included to one solid and excluded from the other. By this procedure parts are determined in form of solids, and finally a combination of parts makes up a product. At the last step a check is performed, does pieces being put together give the product.

Interpretation of a drawing is made by means of analysis of space relationships among the main entities with respect to the influence of auxiliary ones in the following order: search for break-lines; revealing of hidden lines; recognition of surfaces; recognition of volumes; determination of volumes' status and composition of solids; grouping volumes into parts. Recent minimal contents of the ADI knowledgebase (rules 3-13 for surfaces, 14-22 for volumes, 23-28 for solids, 29-34 for parts) appears to be sufficient to obtain correct results from analysis of real drawings. For example, the case study assembly drawing consists of 82 main (drawn by sick lines) entities, including 6 circles, 19 arcs and 57 lines in two layers: *FRONT* and *TOP*.

When any rule is applied, results are asserted to the database as facts. Mathematical 3-D model (space image) of a product and its parts obtained as a result of Assembly Drawing Interpretation is composed of the following set of facts stored at the database:

(1) *s(Surface#,Surface_type={pln|cyl|con|hex},[X0,Y0,Z0,Tan],[List_of_lines])* - describes equation of the surface with number #: parameters indicate 3-D coordinates of the specific point of the surface and tangent of its cone, or radius of cylinder, or displacement of plane according to:

   cone (at the FRONT view): *Side*((X-X0)**2+(Z-Z0)**2-Tan*(Y-Y0)**2)<0*

   cylinder, hexagon (FRONT view): *Side*((X-X0)**2+(Z-Z0)**2-R**2)<0, R=Tan, Y0=0,*

   plane: *Side*(A*X+B*Y+C*Z+D)<0, A=X0, B=Y0, C=Z0, D=Tan.*

(2) *v(Volume#,Volume_type={cyl|con|hex|box},[List_of_surfaces],Status)* - indicates type of volume with number #; list of surfaces ('+' or '-' indicates "solid" side for each according to the value of *Side* parameter in mathematical description of a surface: *Side*(Ax+By+Cz+D)>0)*); and the status of the volume (1-include, 2-exclude, 3-both; sign '-' means "only", i.e. indicates finally accepted and certain status);

(3) *part(Part#,[List_of_volumes])* - gives information on what volumes from the list of volumes compose each part;

2. Reasoning about Product's Kinematics from Assembly Drawing

When complete 3-D mathematical models of a product and each of its parts are available, it is easy to use them to find out how parts influence each other's freedom of motions To reach this it is enough to shift the "space image" of one part when the other is stoned still. As was shown in earlier works of authors, this information is sufficient for determination of kinematics and assembly sequences for the whole product. Hence results of assembly drawing interpretation can be applied to robotic assembly process planning. Having the number of parts and their space images, program tries to fond out a set of kinematic relations

$$r(I,J,Dir=\{+X,-X,+Y,-Y,+Z,-Z\})$$

between different parts *I* and *J*, where *Dir* indicates directions along main coordinates, in which part *I* can be moved if all other parts except *J* are taken away. Results of this analysis are presented as a table where marked fields mean possibility to move part *I* relatively to part *J* in the coordinate direction assigned to the column. Here information obtained at the previous step can be checked and corrected and then used as input information for automatic determination of assembly sequence alternatives. Program available at the recent research stage allows selection in dialogue for optimal sequence of assembly through disassembly procedure. By analysis of input information program consequently affords for expert analysis and decision all acceptable by product's design disassembly actions. The sequence to be formed is depicted dynamically as a string of parts numbers, where subassemblies are distinguished by brackets. The result of assembly sequencing may be saved to disk with all information for backtracking. Program mensioned makes possible sequencing of assembly for the most complicated products with number of parts exceeding 12 (in many cases even more).

176

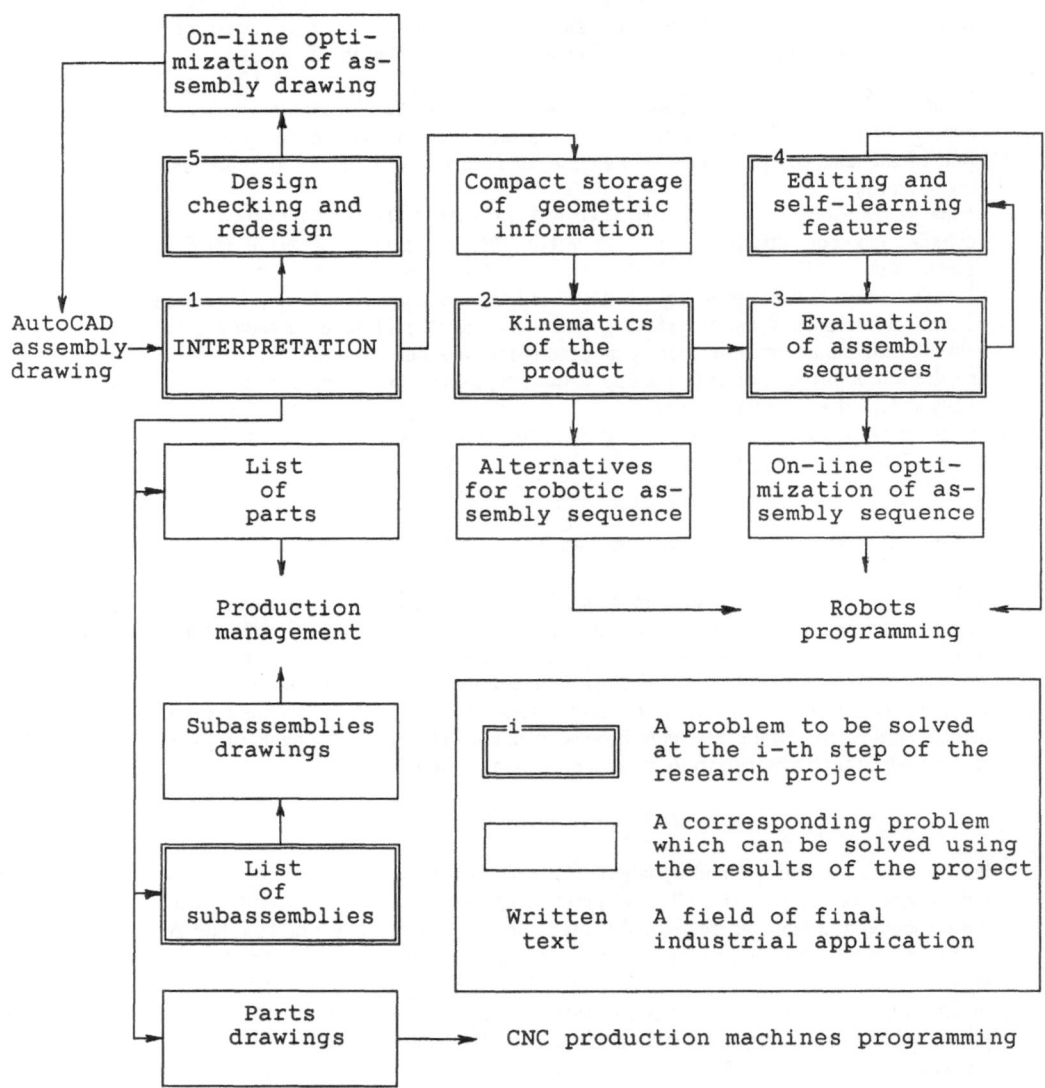

Fig.1. Reasoning about Assembly Drawings: research project
fields of application

3. Advanced Applications

Models obtained as a result of assembly drawing interpretation give detailed and complete mathematical description of the product as a set of parts specifically arranged in modelling space: from the presentation convenient for perception by user a description of a product is transformed into the form suitable for modelling of design and technological improvements of a product on PC.

A System for Reasoning about Assembly Drawings (ADI) is the main part of a complex research project (Fig.1). Mathematical model of a product, some rules and methods of its construction described in this paper are subjects for at least three further research ways: automatic generation of draft drawings of parts with interface to automatic programming of CNC production machines; optimization of assembly process for ensuring complete utilization of specific design-for-assembly features introduced into the product during its design; assembly production knowledge support for user-designer with prospects towards automaic redesign for better robotic assembly. Since any knowledge need updating and support, some Knowledge Base Editing System (with self-learning features preferably) would be very helpful. At Fig.1 this feature is shown as a support tool for Assembly Technology Evaluation task.

Automatic generation of a set of parts draft-drawings through interpretation of assembly drawing being an important time-saving practical task may be achieved as soon as mathematical models of parts are obtained and checked, with the help of Artificial Intelligence system able to apply rules about how to make parts drawings. The same way reasoning about subassemblies and their drawings can be implemented with automatic subassemblies draft-drawings generation.

Evaluation of Assembly Sequences for better Assembly Production Technology is another important application of research. Product's kinematics analysis described in this paper can be successfully applied here with little additional interface: alternatives must be arranged and proposed as preferable solutions to user in strict order according to their priorities. Further development of this direction is aimed at robotic assembly production planning.

Design Checking and Redesign are important features of Design for Assembly. Though there are already software tools which try to solve this task each in their own way, some new approach based on Drawings Understanding could be helpful. The system considered could not only check and estimate assembly drawing in real time, but also propose some solutions as ready-to-insert blocks to the drawing. Together with components mentioned this could make up a powerful time-saving system for industrial applications of robots and CNC machines.

Acknowledgement

Authors express their words of thanks to o.Univ.-Prof. Dr. P.Kopacek from Institute of Robotics at Technical University of Vienna for his support in realization of this project.

Scheduling in Flexible Manufacturing Systems, Supported through Cyclic Net Analysis

L. Lenart, A. Ruzic
Jozef Stefan Institute, Robotics Laboratory
Ljubljana, Slovenia

Abstract: The analytical solutions of scheduling problems in flexible manufacturing systems (FMS) can seldom be used in real time scheduling as their numerical complexity is in the NP class. However the experience with analytical solutions based on relatively small objects can be often useful when constructing real time algorithm for some basic scheduling algorithms in FMS. In problems handled, the set up mechanisms are not neglected.

1   Introduction

Often the set up problem in FMS models can be overviewed, see general reports (E.L.Lawler, 1982) or (S.Greves, 1981)), the recent research in real time control (J.R.Perkins, 1989) recovers it again. It is intended in this paper to treat the set up problem in classical scheduling algorithms and to propose some real time algorithms in such environment.

In the second paragraph, the single machine set up problem is treated, the exact and simple real time algorithms are compared. The set up problem is treated in a multi-machine environment under 'just in time' (JIT) synchronisation condition in the third paragraph. The flow shop problem is handled with set up ability, the solutions for normal and long term schedule are compared.

2   Single Machine Scheduling with set-up

The FMS systems are often characterised through the possibility to set up the working machines in the time which is comparable with the time to produce limited number of workpieces. Over the properly long production interval the optimal policy then should optimize the criterion function, this is normally the shortest makespan of the planned production with given product mix. As first, the problem of finding the optimal policy can be embedded in the stochastic or in the deterministic environment. The former case leads to the stochastic decision problems and the later to the deterministic scheduling. Here we will point out only the deterministic problems.

Single machine set up problem

The problem of machine set-up in his most simple form is presented as example (J.M.Korsunov, 1980) , allready indicating some questions which are common for set up in much more general FMS tasks. Let the final product be produced on a single machine in n operations. Each operation requires the processing time l_i; $i = 1, \ldots, n$. The sequence of operations is not prescribed, the machine can execute them in any order. They are $n!$ possible sequences to finish the work and the machine busy time MBT is allways $\sum_{i=1}^{n} l_i$. After the i-th operation the machine must be set up for the k-th operation, what can be done in $d_{i,k}$ time units. The operation schedule must be found which minimizes the production time. As the MBT times contribute to the production time with a constant, only the sum of set-up times must be minimized over the set of all possible schedules. The problem now can be formulated as the graph- oriented one, the Hamiltonian path or Hamiltonian circuit must be found.

Numerical example with real time algorithm

Let the graph incidence matrix A be an 8×8 matrix, generated with a Pascal random generator of integer numbers in interval 0 through 9. As the numeric performance of NP algorithm in real time is questionable, suboptimal algorithm should be used. The idea of greedy algorithm can be exploited which suggest to schedule first the most 'perspective' item. Such an algorithm can be written down in the next steps:

1) clear the list of operations L
2) insert to L index i from A with minimal $d_{i,j}$
3) search in the j-th row the minimal d_{j,k_j}, insert j into L
4) change k against j and repeat steps 3,4 until list L is full

The results for shortest Hamiltonian path(circuit) are 8(12) time units, the optimal operation sequences are then : (7,3,8,5,6,2,4,1) resp. (8,5,1,4,7,3,6,2). The number of identical solutions is 3(16). The RT algorithm delivers for path(circuit) the values 13(16) with the operation sequence (5,1,2,8,6,3,4,7), which is the same for path and circuit. The RT algorithm can be estimated quite good comparing the results with the mean value for all combinations, which equals 30.7 time units.

It is evident, that the introduction of single buffer with a capacity c, makes the system more efficient. Define d_{HC} is the length of the shortest Hamilton circuit. The makespan to finish all working phases of a single part is given through expression:

$$PR = \sum_{i=1}^{n} l_i + \frac{d_{HC}}{c} \tag{1}$$

3 Periodic Flowshop Scheduling with Zero Buffers and Machine set up

In the set up problem in the section 2. each working operation follows its predecessor after the set up period has expired. The logical expansion of this problem is to search for multiple machine configuration, where jobs are not waiting at all, neither the predecessor part to be processed nor the machine to be reconfigured. Shortly, all happens 'just in time', no waiting periods and no buffers are required. The problem described follows the lines given in (M.Gondran, 1984) and (D.A.Wismer, 1972) and is completed through the analysis of machine set up problem and periodic functioning.

Assume that n jobs (products) must be finished on m machines. The set of jobs is $(J_1, J_2, \ldots, J_n$. If J_j is further replaced through index j, then performing a job j consists of using successivily a certain number m_j of machines, say $S_j(1), \ldots, S_j(m_j)$, in this order. The same machine may be used more then once for dealing with one piece, but not all machines need be used. The times of working each job j on each machine $S_j(k)$; $k = 1, \ldots, m_j$ are known: they equal τ_{jk}. It is required to schedule the jobs so as to minimize the total time of finishing the n jobs, under the following assumptions:
- Each machine can deal with one job at a time.
- No intermediate storage is possible.
- No job can overtake another during processing.
- When the job number is changed on any machine $S_{j_1}(k)$ from j_1 to j_2 then the set up time is $\delta(j_1, j_2, S_{j_1}(k))$

Now the simple, even time consuming formulas development can be omitted and the final result is presented in the Gantt chart (fig.1). From fig.1 one can see, that task y is delayed for 4 types of summands, $\delta, d_{xy}^1, d_{xy}^2, and d_{xy}^3$. δ is the maximum set up time between any two machines. d_{xy}^1 is the delay of first task of job y, which must be processed on machine M_1. Summands d_{xy}^2 and d_{xy}^3 can be obtained in the analog manner for other tasks.

Extending the fig(1) for other jobs, in the next phase the job y can be taken as reference for the next task z to get the value of d_{yz}. Continuing stepwise, one can get the matrix d_{ij} for all tasks. The problem thus is formally the one of traveling salesman. The total processing time P has to be minimized by choosing the proper sequence $[i_1, i_2, \ldots, i_n]$ of jobs:

$$P = \min_{i_j}\{\sum_{j=1}^{n-1} d_{i,i_j+1} + R_{i_n}\} \tag{2}$$

The problem (2.5) can be reduced to the shortest Hamiltonian path algoritm (Wiesner1972) and solved with proper algorithm.

The secundary result is , that the set up times do not influence the basic algorithm. Evidently, the lack of buffer places where the workpieces could be stored to escape the set up of machines before each working operation effects the productivity in the same sense as found in eq.(1).

real time scheduling algorithm

Let us return again to the case that the technological reasons force the minimal part set (MPS) optimal scheduling is technically optimal too, for instance to keep production inventory limited. The suboptimal real time algorithm should be found. It is an intrinsic property of real time scheduling algorithms, that they are able to make a decision considering only the restricted area of schedule net. The schedule controller can as in paragraph 2. use the following algorithm for one scheduling step:

1) Consider the flow of jobs, that are currently installed on machines
2) Calculate the d_{xy} values for all the others jobs
3) As soon as any job terminates its current operation
   insert the new job with the smallest d_{xy} value.

Furtheron the model in fig.1 can be transformed to Petri net , to the subspecies of it called event graph. With calculus of net invariants all cycles in the model can be found. It is the known result for event graphs that the total number of tokens is invariant by transition firings. For any elementary circuit γ, we can then compute the cycle time:

$$C(\gamma) = \frac{\mu(\gamma)}{M(\gamma))} \qquad (3)$$

where $\mu(\gamma)$ denotes the sum of transition firing times in circuit and $M(\gamma)$ the number of tokens circulating in this circuit. Next we can define π as the maximum cycle time taken over all elementary circuits, i.e. :

$$\pi = \max_{\gamma}(C(\gamma)) \qquad (4)$$

The maximum cycle time determines the system effectivness.

4   Conclusions

The schedule in simple FMS - system was observed, beginning with one machine set up problem and continuing with zero buffer schedule, expanding finally to timed Petri net model. The next facts could be observed or deduced :

- In the one machine scheduling problem in chapter 1. the statistics shows, that quite good approximations to good schedule can be expected even if very simple real time algorithms are used. This suggests the conclusion, that the 'minimax' criteria are useful also for generalised systems.

- In chapter 3. the flow-shop problem with zero buffers and set up was treated. The problem could be reduced to a known classic problem without set up.

- The flow-shop FMS can be observed in the periodical environment. The local optimal schedule is not optimal when considering the makespan and machine workload over longer intervals of time. The real time schedule control algorithm was suggested.

- It was shown, that the Petri nets models are extremly suitable to handle FMS models of the form, as they occur in this contribution.

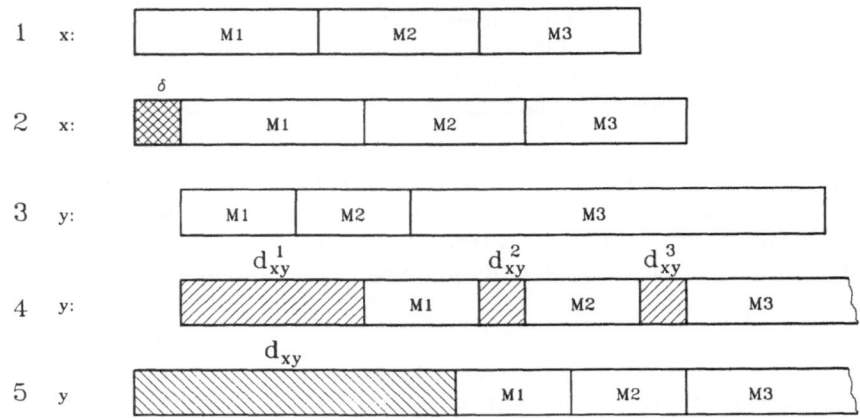

Legend to fig.1:
x .. job x
y .. job y
crosshatched area .. $\max_k \delta_{xyk}$
righthatched area .. d^k_{xy}
lefthatched area .. d_{xy}

Figure 1: Gantt chart for set up problem with zero queue length

References

D.A.Wismer (1972). Solution of the flowshop scheduling problem with no intermediate queus. *Ops.Res*, 20:689 – 697.

E.L.Lawler, J.K.Lenstra, G. K. (1982). *Recent developments in deterministic sequencing and scheduling.* M.A.H.Dempster et al.,Eds. D.Reidel Publishing Comp, Boston,MA.

H.P.Hillion, J. (1989). Performance evaluation of job shop systems using time event graphs. *IEEE Trans. on Automatic Control*, 34(1):3 – 9.

J.M.Korsunov (1980). *Matematiceskie osnovi kybernetiki.* Izd. ENERGIJA, Moskva.

J.R.Perkins, P. (1989). Stable,distributed,real-time scheduling of flexible manufacturing assembly disassembly systems. *IEEE Trans.Aut.Contr.*, 34(2):139 – 147.

M.Gondran, M. (1984). *Graphs and Algorithms.* John Wiley , Inc. NewYork.

S.Greves (1981). A review of production scheduling. *Opns.Res.*, 29(4):646–675.

Computer Aided Planning of Robotized Assembly Systems

M. Skubic
Laboratory for Handling and Assembly Systems Automation,
Faculty of Mechanical Engineering, University of Ljubljana, Slovenia.
D. Noe
Head of the Laboratory for Handling and Assembly Systems Automation,
Faculty of Mechanical Engineering, University of Ljubljana, Slovenia.

Abstract: The process of choosing a suitable robot for an assembly application is time consuming and difficult. If every potential buyer or user has to go through the same time consuming process of collecting specifications and characteristics on numerous industrial robot models suitable for assembly, having to analyze them and make decisions, this is a waste of a large amount of resources to the industry as a whole. To eliminate the above problems a computer database was established to maintain the specification and characteristics information of the assembly robots available on the market. An interactive computer program was developed to carry out the process of search and elimination to help users to select a suitable robot. Incorporated within the selection program, the computer graphics design is used to build and optimize a layout of an assembly station.

1. Introduction

Assembly automation is one of the most promising areas of robots application in the future. The planning and the development of such robotized assembly systems is today carried out mostly manually. A great waste of time at manual planning is one of the barriers in making an offer or an analysis of the corresponding economic variant solution. Successful applications of robotized assembly systems in production dictate the development of a computer aided systems in order to support the planner in building the optimally configured assembly system at low time consumption. Assembly system's components selection is one of the important parts of building a new or modifying an existing assembly system. The degree of risk in the robot selection procedure within the computer aided assembly system planning process is directly related to the unknown decision making parameters, the robot information given by the manufacturer and also the skill of the planner. To eliminate the above mentioned problems a system for computer aided assembly system planning was developed (Noe and Kopacek, 1992). A part of this system is also the robot selection process. Optimal solutions require a systematical determination of selection and decision criteria, creation of an appropriate database, building of a program for computer aided robot selection and at the end the planning system verification process.

The planning system itself consists of two major components. One is methodology which guides the planner from the planning task to the solution. The other important part consists of suitable tools for the planning process. The planning procedure could be divided into several steps as described in (Spur et al., 1985). To create user programs for planning of assembly systems by industrial robots, geometric, technological and economic data are required. Geometric data are available from the CAD system, other data are stored in different databases and managed with the common database management system (DBMS). Development of the planning system is based on the supposition that a flexible assembly system is described as a group of modules, such as robots, grippers, transportation, orientation and ordering devices, various assembly tools, sensors and control that work simultaneously in certain assembly operations. Different planning methods for assembly system design are known and planners are supported by various tools (Bullinger, 1986). In the development of our system these methods were surveyed and taken into consideration.

In spite of a great number of different assembly robots for building the assembly cell, it is possible to systematize the selection and design processes and to create an appropriate robot database. In comparison with other computer aided assembly planning systems (Eversheim, 1989; Feldmann and Hemberger, 1987), our system for assembly robot selection takes into account different robots presented on the market. Information and knowledge are collected from assembly system planning experts, catalogues and manuals. The knowledge is then reduced to the final choice which is made by the planner.

2. Selection of industrial robots for assembly

The purpose of the presented work is to construct a computer database that stores as detailed as possible the specifications information and special characteristics of all the industrial robots available on the market. The data of each robot and selection process are accompanied by an interactive computer layout design program. In research work attention was focused on design of a user-friendly interface. The database servicing (adding, updating and deleting of records) and selection procedure are so written that users do not have to know a lot about the relational database, its structure and relations between entities. If an error occurs, the program writes a warning message and the user can correct it so there is no need to start the procedure from the beginning. In addition, a graphics library of robots was made and an interface between the selection program output and the graphic system was written. In the future the system will offer a possibility to optimize the layout of the robot and peripheral devices for minimizing assembly time and assembly cell costs.

All the collected information is stored and maintained in a database that can be accessed at very high speed by computer. A controlled approach is used in adding new data and modifying and retrieving existing data. The major components of the industrial robot database covering the specifications and characteristics information described in the previous section are illustrated in Figure 1. It shows that we could divide robot's data in different groups like function data, geometrical data, etc. The robot database is a part of the planning software package and is managed by the database management system. The planning database contains also data for different devices of the robot system. These are the basic robot, robot control, grippers and sensors. The developed program assists the user in selecting suitable assembly robots without going through the difficult and time consuming process described above. What the user needs to do is to answer a series of questions asked by the computer and the computer carries out the process of search and elimination. A part of robot selection program structure is shown in Figure 2. It can be expanded to include other robot characteristics menu levels along the dotted-line path. At the end of the process, the computer lists the candidate robots that meet the requirement specified by the user.

Having selected one or a few industrial robots that possess all the required specifications and characteristics, the user very often wonder whether the robot can be incorporated into existing shop-floor or assembly cell layout and perform expected tasks or operations when installed. As the second part of this project to put this concern aside, a graphics library of assembly robots and interactive layout design of a robotized assembly system will be described in the next paragraphs.

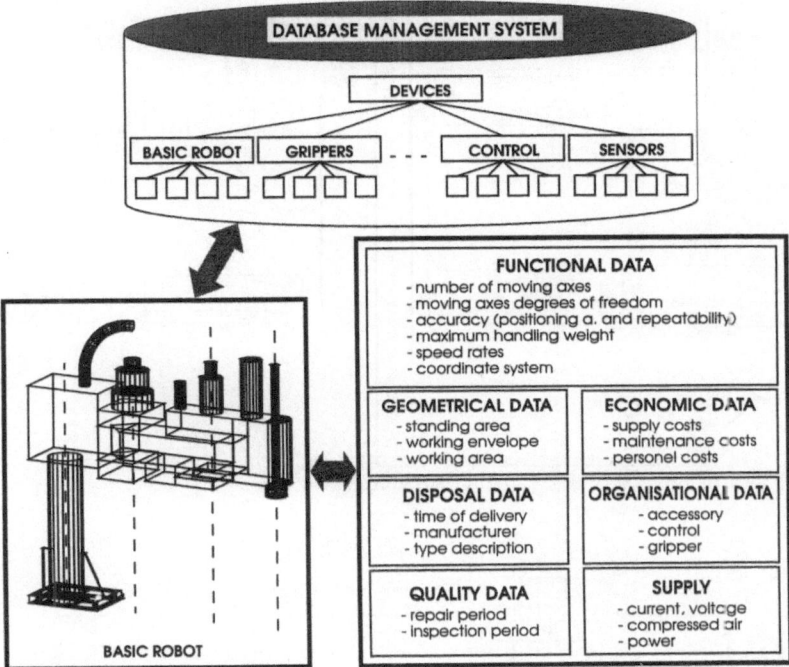

Fig. 1: The industrial robot database

Due to strong emphasis on graphical and visual aspects during the planning procedure, it is appropriate to start with the available CAD system. Based on the three-dimensional design system AutoCAD, a graphical computer database was created. The graphical models of assembly robots are made of surface elements. Such a model is represented as a hollow object with an infinitely thin shell that corresponds to the object shape forms its physical appearance and also the geometrical database. For interactive creation of models some programs were written in the programming language AutoLISP that simplify the work, and speed up repetitive processes. A great advantage of the system for CAD is that the drawing contains also non-geometrical data and could be connected with DBMS. A program written in AutoLISP inserts information into the drawing from a database file. It reads data every time the drawing is loaded into the graphics editor, so they are always up-to-date.

The structure of the entire planning software package is shown in Figure 3. Within AutoCAD the planner runs a program for selection of a suitable robot, leaves AutoCAD environment, and in program selects the robot, based on the criteria and limitations mentioned in previous sections. The next step is exit from DBMS and return to CAD system. By running an appropriate command the selected robot model is inserted into the drawing and incorporated into the assembly system layout.

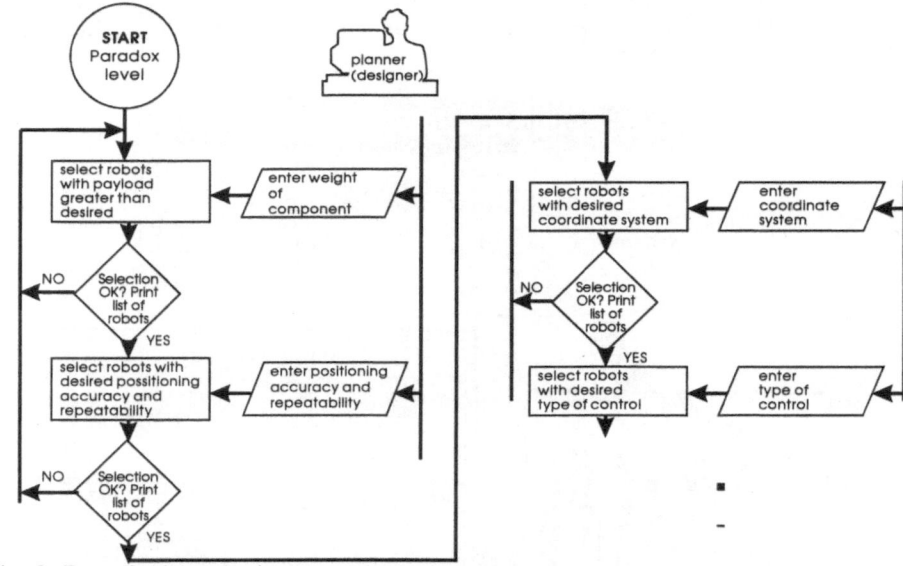

Fig. 2: Part of robot selection program structure

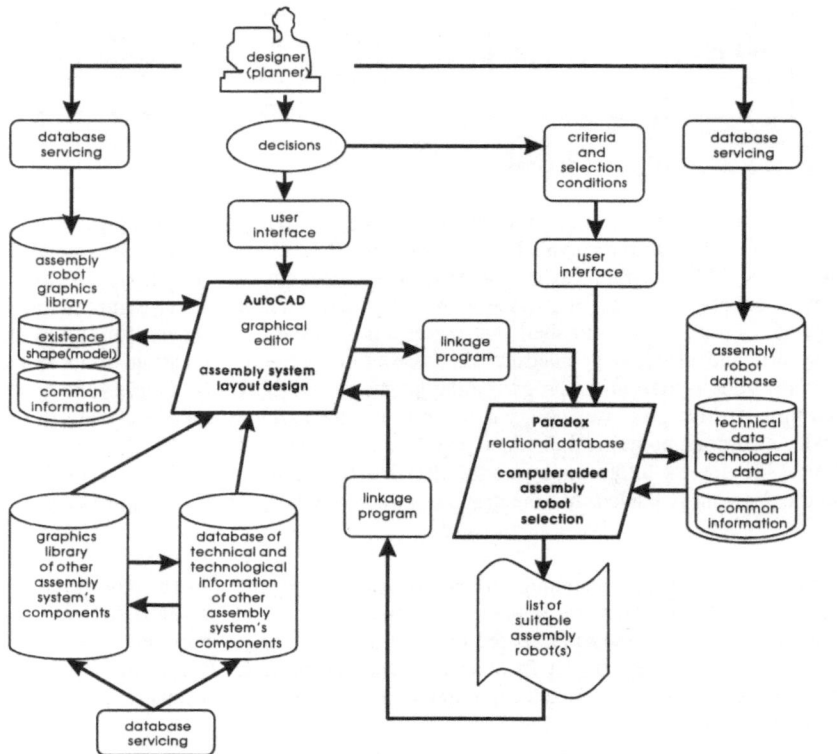

Fig. 3: The structure of selection software package

At a later stage we plan to carry out also an economic analysis making the final decision in an additional module of the planning software.

3. Conclusions

Industrial robots have proved to be a key technology for flexible automation in assembly. Companies will have to intensify their automation efforts to maintain the leading edge in production technology. However, the current obstacles which influence the planning of robot integrated assembly systems have to be overcome in order to reach this objective.

Therefore an interactive program was written to assist users in selecting suitable robot for a desired assembly task. The graphics library of assembly robots was also built up. The technique of computer graphics design was incorporated into the system to ensure the chosen robot will meet assembly system layout demands and could perform the tasks. The application of computer tools for planning purposes offers the following advantages:

- simplification of selection process,
- increase in planning quality,
- reduction of the planning time and costs,
- consideration of alternative solutions,
- reduction of routine manual work and increase in the share of creative human work,
- greater productivity and efficiency of design,
- improvement of optical presentation,
- possibility of 3D view.

4. References

Bullinger, H.J. (1986). Systematische Montageplanung. Carl Hanser Verlag, München.

Daßler, R., Schafer, G. (1987). CAD und Datenbank zur Planung flexibler Montageanlagen. ZwF, Vol. 10, No. 10, pp. 575-578.

Eversheim, W. (1989). Expertensysteme und Simulation zur Planung von Montagesystemen, Durchlauzeit reduzieren. Technische Rundschau No. 23, pp. 24-29.

Feldmann, K., Hemberger, A. (1987). Rechnereinsatz in der Montageplanung. VDI-Z, Vol. 129 No. 5, pp. 76-81.

Noe, D., Kopacek, P. (1992). A Computer Aided System for Planning Flexible Assembly Cells. IFAC Workshop on A Cost Effective Use of Computer Aided Technologies and Integration Methods in Small and Medium Sized Companies (CIM 92), Wien.

Spur, G., et. al. (1985). Design Rules for Integration of Industrial Robots into CIM-Systems. ESPRIT '84. Amsterdam, North Holland Publishing Company.

Tönshoff, H.K., Menzel, E., Park, H.S. (1992): A Knowledge-Based System for Automated Assembly Planning. Annals of the CIRP, Vol. 41 No.1, pp. 19-24.

Van Brussel, H. (1990). Planning and Scheduling of Assembly Systems. Annals of the CIRP, Vol. 39 No. 2, pp. 637-644.

A user-friendly Software Tool for the semiautomatic Design of small Assembly Cells

R. Kratschmann
Department of Systems Engineering and Automation
Scientific Academy of Lower Austria

P. Havlicek
Faculty of Informatic
Technical University of Brno

Abstract: Today the characterization 'productivity' requires a strictly cost optimized generation of offers. Small assembly cells with at least one and maximum two robots allow flexible assembly steps to produce a broad variety of products. Starting from these requirements, a new tool for the semiautomatic generation for assembly cell offers was developed. Therefore, a design tool must be very flexible to consider all possible product types. New graphic user interfaces (MS-Windows, IBM OS/2, . . .) allow the designer the easy generation of such user-friendly software tools to fulfil all requirements of the client/server architecture.

1. Introduction

An optimal, timesaving work on offers reduces the non-productive time of every firm. This reduction is very important all time and especially during times of economic recessions. The design of small assembly cells with a maximum of two robots is a typical recess for small and medium sized companies. They must react fast and flexible to win most of the requested offers.

The layout of the assembly cell will be designed with the semiautomatic and database supported Planning Software. It is possible to compute the cyclic time and the price of the assembly cell as well as the maximum delivery time of the components. Additionally, it is possible to generate an export file for the AutoCAD-based simulation software SITAR.

2. Application Structure

Personal Computers are cheap alternatives to standard workstations. The falling hard- and software prices and the growing software capabilities allow the use of Graphical User Interfaces (GUI) during the software development. MS-Windows (Version 3.1) is the best example for PC-GUI's, with a large amount of existing applications and development packages. The Planning Software requires a relational database to store the total input and output information that is necessary for the planning progress. The selection of SQLWindows as a MS-Windows database frontend development tool in connection with the SQL database backend of ORACLE (Version 6.0) creates a good combination to develop such a PC-GUI-based Planning Software.

2.1. Structure of the Database

Four relational connected database parts in three different layers (Fig. 1) of hierarchy are forming the support database.

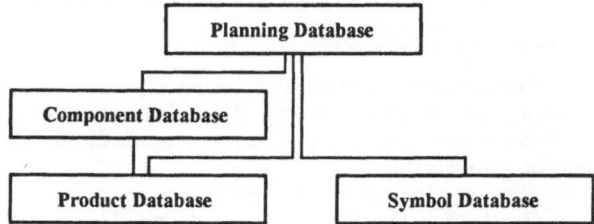

Fig. 1: Hierarchy structure of the supporting database

The 'highest' level of hierarchy will be formed from the Planning Database. It requires data from all other databases. The central layer contains the Component Database and the 'lowest' Level is formed by Product and Symbol Database which data are necessary for all other databases.

Product Database

Main objective of this database part is the storage of general product, manufacturer, reference person and branch data. Each product could be divided into jointing pieces, semifinished products, and auxiliary materials (glue, varnish, . . .). Additional storage of specific product data like general shape or gripping method of jointing pieces and semifinished products can be carried out. These additional information could be used from the Planning Software for an automatic gripper selection.

Data input can be done during the start-up of the Planning Software or from an external database maintenance tool which will be delivered in this software package. Data deletion and modification could be realized only from the external tool.

Symbol Database

The Planning Software is icon driven and each necessary icon data is stored in this database. These data will not be changed during the planning process, a read-only access is done. Data deletion and modification could be done only from an external maintenance tool.

Component Database

Each component which is used from the Planning Software is stored in this treelike database. As the largest of the four database parts it includes general and specific data of a component. General data like prices, supplier, time of delivery or manufacturer should reduce the time-consuming search in catalogues and handbooks. Component specific data is stored in six main- and twelve subgroups. The main group contains general data of robots, grippers, tools, tool changing systems, transport systems (conveyor belt, . . .) and preparations components (storage system, pallet, depot, . . .). The subgroups are mainly used from the Planning Software to select specific component data which is necessary for the planning process. The data must be inserted, deleted, and modified from an external maintenance tool. Without a large number of different component records it is not possible to produce a flexible planning.

Planning Database

Each result which was generated by the interactive Planning Software will be written into the Planning Database. Their entries could be generated only from the Planning Software, an external tools was designed as a viewer to monitor the entries without changes.

2.2. Structure of the Planning Software

Considering the aim that should be reached a user-friendly approach must be found. Every GUI uses object-symbols (icons), so an icon driven solution can be the approach to this problem. Every predefined icon represents one assembly operation or one manufacturing operation like gluing or soldering (Fig. 2).

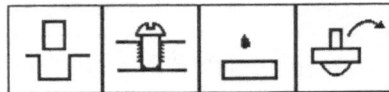

Fig. 2: Examples for assembly or manufacturing symbols: 1) Inserting, 2) Screwing,
3) Gluing, and 4) Handling of an assembly group

In the following description the total planning process will be presented, starting with the entry of new products until the export of a simulation-file.

Description of the product and input of target values

In this module the input of the general product description, definition of the gripping method and the splitting into jointing pieces, semifinished products, and auxiliary materials will be done. All inserts are also possible from an external tool of the product database, so this module can be skipped at the beginning of the work. Additional is the definition of the target values of the assembly cell, namely the cyclic time, price and maximal delivery time of a component.

Description of assembly/manufacturing operations and definition of their order

Every operation of the assembly cell must be described as a list of predefined icons. After the selection of an icon a jointing piece, semifinished product, or auxiliary material must be selected. Also the required number of robots, robot-joints and the number repetitions is necessary (Fig. 3).

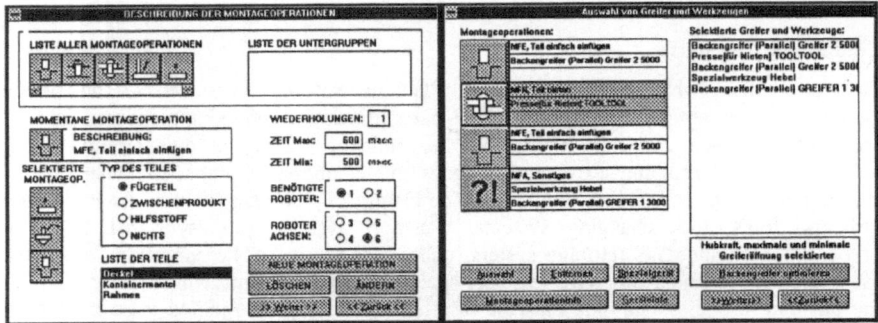

Fig. 3 & 4: Screen hardcopy of the description of assembly/manufacturing operations and of the tool/gripper selection

Splitting of assembly/manufacturing operations

With the stored times of the selected operations a coarse cyclic time could be calculated. If this time is larger than the target value, the use of two robots will be suggested and/or a division into two or more assembly cells is possible. If two robots are selected, it is possible to set waiting-points if one operation must be done before another one. The division into two robots must be done manually because no algorithm was found until now.

Selection of tools and grippers

After the division a gripper must be connected with an assembly operation and a tool with a manufacturing operation. The connection will be preselected from the Planning Software and the user must select an offered result. A gripper is preselected if it could carry the mass of the connected jointing piece or semifinished product and if the connected gripping method is correct. After the selection of all grippers and tools an optimization of the selected grippers can be done from the Planning Software (Fig. 4).

Connection of a tool/gripper with a tool changing plate

The necessity of a tool changing system will be checked and the tools/grippers can be connected with a tool changing plate. The algorithm for the selection of the changing system depends from the number of tools and grippers

Refinement of assembly/manufacturing operations and coarse cyclic time

Each defined operation must be refined with one or more of seven refinement operations. These refinement operations are: gripper or tool feeding, gripper adjusting, gripper or tool changing, part preparation, and part feeding.

As a result of the refinement a precise cyclic time of the assembly cell can be calculated and stored for further analysing.

Input of required components

A semiautomatic selection of tool changing systems, one or two robots, preparation components, or transport systems from the Component Database which depends on the planning process will be supported from the Planning Software. If a specific component was not chosen, additional components can be chosen without any preselection.

Output of results

The first point of the result presentation is the comparison of the calculated price, cyclic time and delivery time with the target values. If the results are sufficient the user must select "accepts results" to create a printout of required components and the comparison with the target values or "drop results" to restart the planning process.

A list of all selected components could be printed on the screen and the printer. The list includes the name of component, distributor, delivery time and price of each selected item.

A file (*.INI-style of MS-Windows) which includes a detailed description of all selected assembly/manufacturing operations and the important robot value represents the output to the SITAR simulation software.

3. Summary

The trend to modern GUI's and a high user friendliness require powerful development tools to create a modern application. A supplement of the user during the planning process of a robotized assembly cell saves a lot of development and offer time. As a disadvantage the user must have a broad knowledge about assembly cells to verify all the results and messages of the Planning Software. Without this knowledge within the automation it is impossible for him to create realistic and reasonable results. Nevertheless, the application presents a powerful tool for future planning of costs and processes in a wide range of cell manufacturing plants.

4. Bibliography

Noe, D., Kopacek, P. (1992). Computer Aided Planning of Flexible, Modular Assembly ells. In: Proceedings of Robotics in Alpe Adria Region. Portoroz, 139 - 147

Kopacek, P., Probst, R. (1992). Assembly of Primary Parts of Welding Transformers in a Flexible, Robotized Cell. In: Proceedings of Robotics in Alpe Adria Region. Portoroz, 148 - 153

Data Structures and Procedures for Computer Representation of Robot Cells

A. Ruzic
Jozef Stefan Institute, Robotics Laboratory
Ljubljana, Slovenia

Abstract: Several programming packages used for research, design, and application in robotics require some capability for computer modelling and representation of different aspects of robots, robot cells and other cell components. In the paper we discuss the requirements and characteristics of such computer representation systems. Then we describe a programming system for robot cell modelling and simulation that we designed primarily for research purposes. We give the overview of the global structure of the system, of its modelling procedures and data structures.

1 Introduction

Various programming systems used in robotics require some capability for the computer modelling and representation of different aspects of robot and robot cell characteristics. The examples of such systems are:

- Robot controllers. Current robot languages tipically offer the possibility to define cooordinate frames attached to different objects and relations between them. This represents a rudimentary modelling of the environment.

- Sensory systems. They acquire low-level information from their physical sensors. The embedded computing subsystem then builds a higher-level and more structured representation of the observed characteristic.

- Off-line programming systems. In these systems some characteristics of the robot and other components involved in the robot's activity are modelled. The systems can simulate the effects of robot program execution on the defined model.

- Robot and robot cell layout design systems. Special systems for the design of the robot's mechanical structure, for the design of its control system and also for the determination of optimal cell layout require adequate models of robots and components.

- Systems for the research and development purposes.

We are interested especially in the last kind of systems. Modelling and representation of robot characteristics and simulation of robot behaviour is especially important in the robotics research. In the first place, systems which support these functions are used as environments for the design and test of the novel control methods. In the second place, they are an inherent and necessary part of some advanced control methods. In manipulator-level robotics, the controller intelligence is mainly responsible and limited to the control of the kinematical and dinamical aspects of the mechanical manipulator. All other functions are implemented by the user. He constructs the robot's ability to cope with a number of foreseen situations by writing an adequate robot program. One of the goals of robotics is to design advanced robots. Their tasks should be defined on a high level and they should be able to adapt autonomously to changes in their working environments. Their respective control systems should incorporate a number of advanced strategies that require models of different characteristics of their environments. E.g., high-level task planning and motion planning require the 3-D models of the robot, positioning devices, transport systems, workpieces, and other components involved in the robotized process.

2 Requirements and characteristics

We developed a robot modelling and simulation system for our research and development purposes. Our system, like other systems of this kind is not meant as an alternative to dedicated commercial robot simulation systems. Some representatives of such marketed systems are Tecnomatix's Robcad (Carter, 1987, Hollingum, 1990), B.Y.G. Systems' Grasp (Bonney and Young, 1987), Deneb Robotics' Igrip (Harrison and Mahajan, 1986) and SILMA's CimStation. These systems have friendly user-interfaces and ready kinematic solutions for a considerable number of industrial robot structures.

However, previously described systems have a number of limitations when used for research and experimental purposes. They usually address modelling of robot's geometry and kinematics. Modelling of other characteristics as communication and sensory capabilities is limited or not possible. Some of them enable simulation of a single robot mechanism at a time. The range of analyses that can be performed is usually fixed or limited. The possibility of other equipment programming is also limited. They offer no possibilities to address the modelling of general equipment, of physical processes that describe interaction between cell components, of uncertainty, of sensory processes and of part processing. The most important issue is that it is difficult or impossible to integrate user-written procedures to perform these functions.

For these purposes, a number of open and research robot modelling and simulation systems have been developed. Some representatives of such systems are RIPE (Miller and Lennox, 1991), RCCL (Hayward and Paul, 1986, Ferrie et.al., 1991), Kali (Backed et.al., 1989, Ferrie et.al., 1991), WADE (Levas and Jayaraman, 1989), ROSI (Schneider, 1991). In respect to its purpose, our system also falls in this category. They does not offer all the possibilities we mentioned before — but they enable the user to integrate them and experiment with them.

Although each of them supports specific experiments, there are common requirements and characteristics for all of them. The creation of models must be reasonably simple, data structures must enable the design of efficient algorithms that use them and the system must be extensible to accommodate different representations of the same aspect.

Their main characteristic is that they are absolutely open. They must namely enable

- easy and effective integration of the system with the user-written procedures,

- performance evaluation of different algorithms,

- modification, extension and refinement of system's components with reasonable, relatively low effort.

The last requirement is common to many programming systems. It is achieved through careful modularisation, use of standard program building blocks etc. Recently, design of object-oriented environments becomes an important method for achieving this goal (Levas and Jayaraman, 1989, Miller and Lennox, 1991).

A number of research modelling and simulation systems enable also on-line connection to target controllers or even on-line manipulator control for the verification of simulated algorithms on real objects. The examples of such systems are Handey (Lozano-Perez et.al., 1989) and the already mentioned RIPE, RCCL and Kali.

As openness and flexibility are the main requirements for such systems, they often lack the user-friendly interface of other systems. Such interfaces are namely designed to improve effectiveness of trained operators when performing restricted sets of standard or permitted operations, but they prevent access to more embedded system's functions.

3 System description

In the following section, we give an overview of the architecture of ROMAS (RObot Modelling And Simulation system), a robot modelling and simulation system that we have developed for our research and development purposes. At the beginning, we wanted to have a system for graphic visualisation of robot movements when using different contol methods. Then we upgraded it and used as a testbed for the development of a robot programming language and for determination of robot cell layouts. Currently, we use it for the research of motion planning and other research purposes. We will describe the main characteristics of its structure an give an overview of its functions.

The first fact one can determine is that it is impossible to design a fixed and predefined structure of data and algorithms for the representation of all characteristics and for simulation purposes. Therefore we *modularised* the system. *Modules* are composed from data structures and procedures performing operations on them.

The purpose of our robot modelling system is to support experiments and implementation of various algorithms. The effectiveness of each algorithm often depends strongly on the type of data structure representing the model of some characteristic. Therefore our second conclusion was to design such robot cell representation scheme which can accomodate *multiple representation schemes* eow for the model of the *same characteristics*. In this way the system enables effective implementation of new algorithms, as each of the algorithms can use the most appropriate representation.

The third characteristic of our system is that it has an unified framework for the representation of different physical entities or devices. The main data structure is the *environment*. It can represent items from the most simple one, e.g. a simple component used in the production process, to the most complex one, e.g. a robotized workcell comprising several robots and positioning devices. Its base element is a component node which can be linked to the other component(s) to form more complex devices and environments. Special procedures for handling of composed devices can be written. Depending on the type of devices included in the environment, data structures and procedures from appropriate modules are involved in the construction of the model and used in the simulation of its behaviour.

The core of the system is realized using Pascal procedures and data structures, grouped into five modules. The first module is the *solid modeler*, which consists of a number of procedures for the following functions: defining primitive solids, translating and rotating solids, composing solids, defining surface characteristics and appearance, retrieving and storing solids' representations. The procedures, implemented as Pascal functions, represent a kind of solid modelling command language. The user defines a solid by writing a source alphanumeric description — a Pascal program — using these functions. From the solid modelling standpoint, the user defines solids using a restricted CSG (constructive solid geometry) scheme with a glue operator (Requicha, 1980, Morteson, 1985). The input domain is represented by polyhedra. The input description is translated to a primary internal CSG description and a secondary boundary representation which are stored in the modeler's database. While the input description represents a readable form which can be easily modified by the user, the internal descriptions are in a form that can be efficiently used by various application algorithms.

The embedded modeler we have developed is used for the fast description of modestly complex solids. We use SDRC's I-DEAS solid modeler for construction of arbitrary solids. We have implemented conversion from the I-DEAS universal file solid description to ROMAS' internal description.

The *kinematic modeler* uses a similar scheme for the definition, translation and storing of kinematic chains. It has functions for: defining general kinematic chains (robots, grippers, positioners etc.) composed by rotational, translational and parallelogram joints, assigning solids

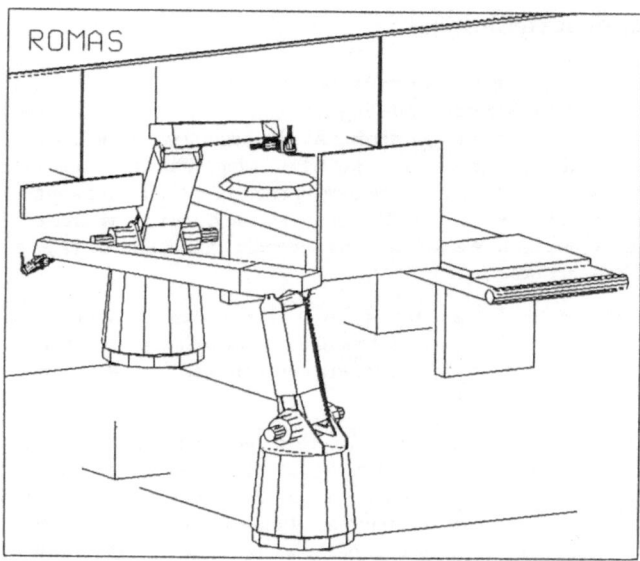

Figure 1: Simulation of robot cell activities in ROMAS.

to degrees of freedom, defining joint attributes, storing and retrieving chain representations. Each robot representation usually contains complete information about the robot's kinematic and geometric structure.

The *cell modeler* implements functions for: creating new cell layouts using previously defined components and cells, defining spatial relationships and type of connections between components, storing and retrieving cell representations.

Using the described solid, kinematic, and cell modeler, the user creates representations of parts, components, robots, devices and cells. These representations are subsequently used in simulation of cell activities. The *simulation group* of functions includes features for: loading cell representations, programming robots, positioners, grippers and other components with joint displacement commands, commands for defining changes in spatial and connection relationships, and procedures for off-line simulation of cell behavior during program execution.

During the simulation, the system calculates the effects of the program execution on the cell structure. The system also performs hidden-line or color shading graphical animation of simulated cell activities, as it is shown on figure 1. To enhance the interface between the user and the system, we are developing an *interface monitor*. Using this monitor, the user accesses the systems' functions through a windowing environment and a command language. Because of its capabilities, ROMAS can be used for choosing appropriate robots for specific tasks, evaluating alternative cell layouts and developing logical and positional correct program structures. For our purposes, the most important characteristic is that the system is open and therefore it enables implementation and embedding of additional features, e.g., inverse kinematic calculation, task-level programming commands and simulation of more complex cell components. It can therefore be used as an environment for the robot software development and robotics research support.

The system is written in Pascal and presently runs under VMS operating system on VAX computers. The current user and graphics interface are written for Tektronix 412x and 423x graphics terminals. To enhance the system's capabilities, to achieve portability between different platforms and to enable connection of the programming and simulation environment with target robot systems, we plan to port the system to an entry level graphics workstation running Unix and using standard libraries and tools for the user and graphics interfaces (X11, OSF/Motif, Phigs).

4   Conclusion

A number of robot programming systems include a subsystem for some kind of computer modelling and representation of robot and robot cell characteristics. In the paper we have given an overview of the modelling and simulation system developed primarily for the research purposes. It provides capabilities for modelling of the geometrical, kinematical and relational characteristics of devices forming a robotized cell. The system is open to include other representations of the same characteristics and other models.

Up to now the system has been used as an environment for the development of a robot programming language, for the simulation of robot cells and to support research of robot motion planning.

References

Backed, P., Hayati, S., Hayward, V., and Tso, K. (1989). The KALI multi-arm robot programming and control environment. Proc. NASA Conf. Space Telerobotics, Jan. 31-Feb. 2.

Bonney, M.C. and Young, K.Y. (1987). Off-line programming using the GRASP robot simulation system. Second International Conference on Computer-Aided Production Engineering, pp. 67-70.

Carter, S. (1987). Off-line programming: the state-of-the-art. The Industrial Robot, vol. 14, no. 4, pp. 213-215.

Ferrie, F., Hayward, V., and Dalziel, M. (1991). Experimental Robotics Research at McRCIM. Robotics and Automation, Newsletter of the IEEE R. & A. Society, September, vol. 5, no. 4, p. 13.

Harrison, C.P. and Mahajan, R. (1986). The IGRIP approach to off-line programming and workcell design. Robotics Today, vol. 8, no. 4, pp. 25-26.

Hayward, V. and Paul, R. (1986). Robot manipulator control under UNIX RCCL: A robot control 'C' library. Int. J. Robot. Res., Winter, vol. 5, pp. 94-111.

Hollingum, J. (1986) Robot simulation system comes to Britain. The Industrial Robot, Winter, vol. 17, no. 4, pp. 181-183.

Levas, A. and Jayaraman, R. (1989). WADE: An Object-Oriented Environment for Modeling and Simulation of Workcell Applications. IEEE Transactions on Robotics and Automation, June, vol. 5, no. 3, pp. 324-336.

Lozano-Perez, T., Jones, J.L., Mazer, E., and O'Donnell, P.A. (1989). Task-Level Planning of Pick-and-Place Robot Motions. *IEEE Comput.*, vol. 22, no. 3, pp. 21-29.

Miller, D.J. and Lennox, R.C. (1991). An Object-Oriented Environment for Robot System Architectures. IEEE Control Systems, February, vol. 11, no. 2, pp. 14-23.

Mortenson, M.E. (1985). Geometric Modeling, New York: John Wiley and Sons.

Requicha, A.A.G. (1992). Representations for Rigid Solids: Theory, Methods, and Systems. Computing Surveys, vol. 12, no. 4, pp. 437-464.

Schneider, S. (1992). CAD/CAM Project. In Rembold, U. and Dillmann, R., Eds., Institute Report, Institute for Real-Time Computer Systems and Robotics, Faculty of Computer Science, University of Karlsruhe, Germany.

Robot Guided Anthropoidic Measuring Device

I. Kovac
Jozef Stefan Institute
Ljubljana, Slovenia
A. Frank
Institute of Production Technology
University of Technology, Graz

Abstract: The paper presents the new anthropoidic measuring device which can move with an industrial robot. After considering the possibilities of introducing an anthropoidic measuring device, we are dealing with the construction and calibration of a measuring device. The main purposes of anthropoidic measuring device are geometrical measuring and robot performance testing. The results of the experiments show that such a system can perform different measuring tasks and that the existing industrial robot ist used as a carrier of the measuring device only.

1. Introduction

Automated quality control is an initial point for achieving the high quality of production by modern flexible manufacturing systems. The industrial robot is usually used to perform differently handling operations. Its basic purpose is to perform different serving tasks exactly and fast. To increase its basic efficiency, we can introduce several additional activities like cleaning, debarring, inspection and so on. From the quality and efficiency point of view, it would be very promising, if the additional performances of geometrical measuring tasks are introduced.

There are some applications, where the conventional industrial robots carry out geometrical measuring tasks, like in the vehicle industry by measuring the bodyshells (*Friedmann, 1986; Schmid, 1987*). At this application, the positions are measured on the bodyshell relative to the same positions on the reference bodyshell. Only relative accuracy is needed here, so the pose repeatability is the most important factor when such a robot be used for measuring tasks. Complex measuring task, where absolute accuracy of an industrial robot is needed the error of relative and absolute position cannot be tolerable anymore. So an existing standard industrial robot, as an autonomous measuring device cannot be satisfactorily introduced.

At existing industrial robots appear various errors, like gear, geometrical, temperature, structure and control errors. Some of these errors can be partly overcome by using the calibration tools (*Kovač and Frank, 92; Duelen and Schrör, 90; Keferstein and Frick, 91*). There are many errors which cannot be carried out to the existing control system. In addition the contribution of random errors causes significant influence to the accuracy of the system.

Because of all these reasons we have developed a new robot suitable anthropoidic measuring device AMG-1. The industrial robot, used in the existing manufacturing system is intended only as a carrier for the new anthropoidic measuring device.

2. Anthropoidic measuring device AMG-1

The main purposes of the anthropoidic measuring device are geometrical measurements and performance testing of the industrial robots (*Kovač, 92*). We deal with a mechanical measuring machine with five rotational axis (Fig.1). The segments are made of light composite materials connected in joints that are made from light Al-alloys. Joints are equipped with precise ball bearings and the most accurate angle measuring systems (encoders). Measuring systems can perform angle position data acquisition with the resolution of 0.18". The new anthropoidic measuring device is planed without its own drive systems. The defined position can be reached by pushing by hand or by industrial robot. To make the moving of an anthropoidic measuring device possible by hand, all horizontal segments have to be statically balanced and compensated with counterweights. If we introduce the anthropoidic measuring device for geometrical measurements, we equip it with a touch trigger probe. After receiving measuring signals from all angle measuring systems, the data are transferred to the mathematical model. It is very important, that the electronic system allows acquisition of all measuring signals simultaneously. The mathematical model takes angle data and calculates the space position in cartesian coordinates.

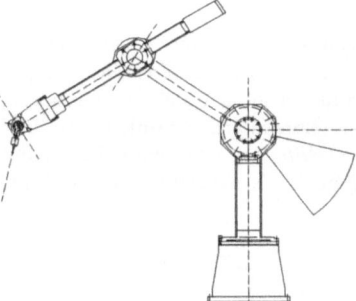

Figure 1: Anthropoidic measuring device AMG-1

By manufacturing and assembling many different errors arise so that the mathematical kinematic model cannot be considered as the real kinematic model. We have developed a mathematical model to consider all geometrical errors and other system and temperature deviations which occur in the anthropoidic measuring device. At the moment, all geometrical errors as parameters of mathematical model of direct kinematics are carried out in the appropriate software.

The preliminary conditions for the precise consideration of geometrical parameters are the accuracy of a measuring (calibrating) machine, where the anthropoidic measuring device AMG-1 is calibrated and the ability of mechanism construction to make the calibration possible. The calibration is done on an accurate coordinate measuring machine, where parameters of geometrical errors and compliance of arm structure of the anthropoidic mechanism are measured. The second condition is the calibration suitable construction (*Kovač and Frank, 92*). For this purpose, we prepared accurate touchable calibers, which are placed on each axis (Fig.2). With these tools it is possible to prepare the measuring device for the medium level of the measuring accuracy.

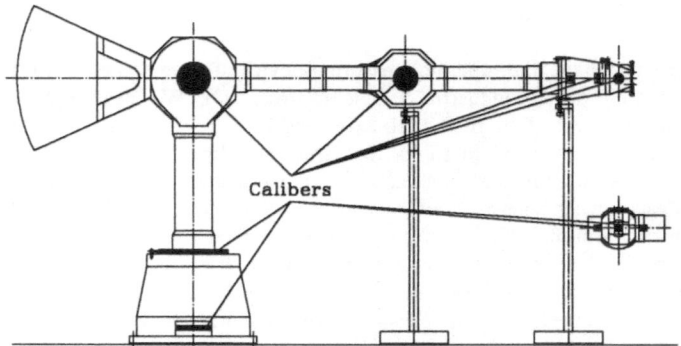

Figure 2: Calibration suitable construction of anthropoidic measuring device

3. Experiments

The measuring accuracy was confirmed by plenty of experiments. First we made experiments without a touch trigger probe. We executed measurements of repeatability, hysteresis, distance and straightness. The repeatability of the anthropoidic measuring device was measured on a precision comparator (Fig.3) in temperature controlled room in the laboratory of Institute of Production Technology in Graz (*Him, 92*). After more than 30 measurements the repeatability of the anthropoidic measuring device is for statistical probability of 95% +/-0.0042 mm.

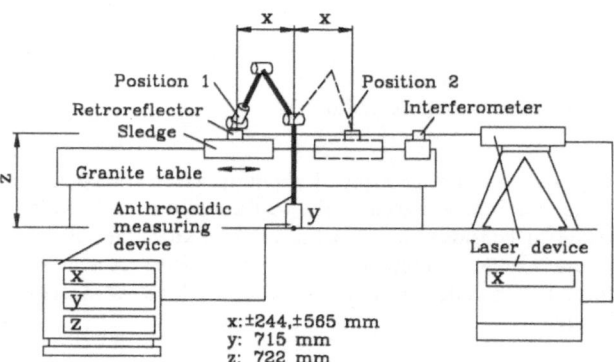

Figure 3: Measuring of repeatability

The accuracy performances were measured on the granite table with precise linear guides and with the laser interferometer (Fig.4). The results presented on the Fig.5 show the distance deviations about x axes before and after compensation. The straightness deviations in y and z direction shows Fig.6.

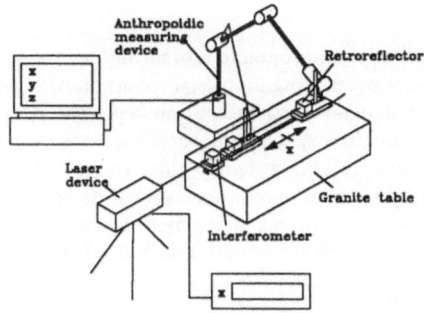

Figure 4: Measuring of accuracy

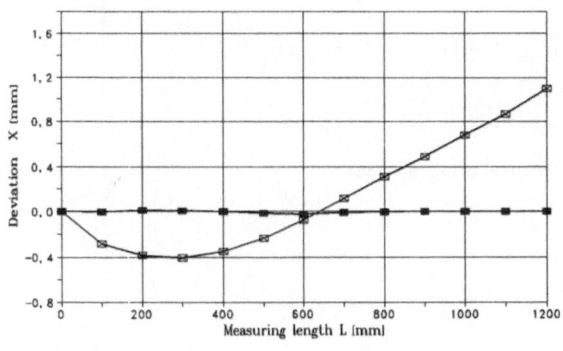

Figure 5: Distance accuracy in X direction

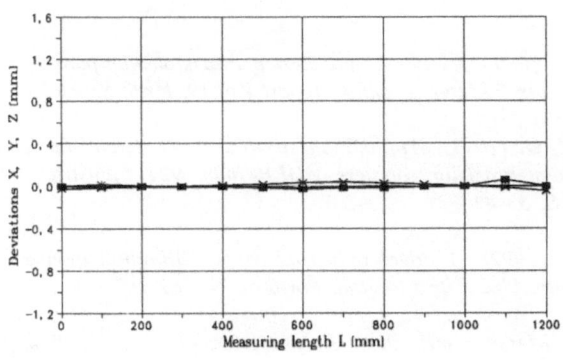

Figure 6: Straightness in Y and Z direction

202

4. Conclusion

In the paper we describe the new anthropoidic measuring device with exclusively rotational axes. The test of performances on existing industrial robots shows that conventional industrial robots cannot be used for the geometrical measuring task. The possibility to overcome this problem is given by introducing the special anthropoidic measuring device which can move with an industrial robot. However the existing industrial robot is by the measuring task involved only as a carrier of an anthropoidic measuring device (Fig.7). The results of the experiments show that such a system can perform different measuring tasks in a flexible manufacturing system equipped with the handling industrial robot.

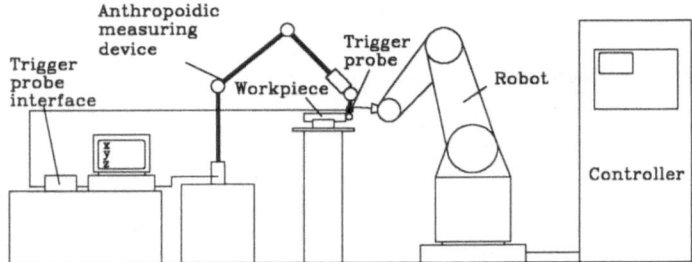

Figure 7: Robot guided measuring device

References

Duelen, G., Schrör, K. (1990). Praktische Resultate der Roboterkalibration, ZwF 85, No-2, S. 113-116

Friedmann, Th. (1986). Roboter in der Automobilindustrie; Robotersysteme 2, Springer Verlag, S. 111-119

Hirn, Ch. (1992). Entwicklung eines klimatisierten Präzisionskomparators für Längen und Geradheitsmessungen bis 5 Meter. Technik Report Vol.19, Nr.6, S. 23-26

Keferstein, C., Frick, O. (1991). Applikationsspezifiscer Abgleich und Kalibrierung eines sechsachsigen Knickarm-Präzisionsroboters. VDI Berichte 921, Industrieroboter messen und prüfen, VDI Verlag, S. 33-42

Kovač, I., Frank, A. (1992). A novel industrial robot calibration device. 1. International Meeting on Robotics in Alpe Adria Region, Portorož, S. 162-167

Kovač, I. (1992). Messen mit dem Industrieroboter - Entwicklung eines neuen anthropoidischen Messgeraetes. Dissertation TU Graz

Schmid, D. (1987). Relativmessungen an Karosserien mit Industrieroboter, VDI/GMA Bericht 14, Industrieroboter messen und prüfen. VDI Verlag, S. 211-223

Circular Test: A new method for testing Industrial Robots

F. Haas and A. Frank
Department of Production Engineering
Graz University of Technology

Abstract: At the Department of Production Engineering at Graz University of Technology we developed a new, fast and effective system for testing industrial robots. It enables the detection of dynamic position deviations caused by robot mechanics as well as the control system and can be applied to the robot under work. In this contribution we want to explain the test procedure and to introduce the method of evaluation. Also the use of a CAD-system to improve the measuring device and especially the possibilities of mechanism design to simulate movements and deviations are described.

1. Introduction

Quality assurance in the production process is becoming more and more important because only high quality products are accepted by the consumers and enable economic success. In addition international standards, especially ISO 9000, recommend the documentation of quality work and demand a periodic control of geometrical machine accuracy. The technical progress, which has been occured during the last years, allows the realization of sophisticated control systems to achieve a constant quality level. A further aspect is the demand to prefer low cost technologies because of worldwide economic problems. Consequently measuring systems are required, which provide real results of high precision in a rather short time without exploding costs. When it is intended to choose an industrial robot for defined applications, it is necessary to compare the specifications of different manufacturers. The experience shows that producers of industrial robots use different testing procedures to determine their product specifications. That is why new standardized testing methods are necessary to compare the deviations of industrial robots. The ISO 230 standard, which defines the suitable measuring instruments, will include the Circular Test to inspect machine tools. In this contribution we want to explain the application of the Circular Test to check industrial robots and the method of evaluation.

2. Circular Test for Industrial Robots

The Circular Test is the measurement principle that includes the comparison between the real circular path of a machine tool and an exact circle, represented by a mechanical reference circle or a high precision bearing. Basically there are two different ways of detecting deviations, first by using sensors which are in contact with the tool centre point and second by applying touchless detectors. If the command path is a circle, the test of path accuracy and repeatability can be carried out by four different methods (Knapp, 1992).

- The first possibility uses a one dimensional tracer which describes a precise circular path. The deviations in relation to a cylinder or sphere fixed at the mechanical interface are detected in connection with the angle position measured by a rotary transducer.
- A circular master piece and a two dimensional probe are the mechanic components of the second principle. The difference between nominal and actual path represents the tracer displacement (Knapp, 1986).
- The third measurement system named double ball bar consists of two precision steelballs which are connected with a telescopic rod equipped with a linear measuring system. The changes in distance are detected and the computer calculates test characteristics.
- A new principle being still in the experimentation stage uses a measuring plate and two scanning heads. The plate, which can be seen as a cross grating, consists of many square rising marks. The main advantage is that any command pathes can be examined by the comparison of real and nominal positions.

At the Department of Production Engineering at Graz University of Technology we developed a new, fast and effective system for testing industrial robots based on the principle of the rotating one dimensional tracer (Fig. 1). It enables detection of dynamic position deviations caused by robot mechanics as well as the control system and can be applied to the robot under work. As the graphic representations of roundness measurements are similar to the results of Industrial Robot Circular Tests, it is obvious to use the software of conventional roundness measurement devices for evaluation. The analogous voltage signals caused by tracer displacements are digitized and related to binary informations produced by the rotary transducer. Both polar diagrams and linear figures illustrate the results because deviations are detected not exclusively in radial direction. The tracer arrangement on a radial movable arm enables high flexibility concerning the command path radius. So the test conditions can be adapted to the individual robot task within the work space. Further a robust construction and an easy assembling on an adjustable stand are basic requirements to carry out tests in industrial environments without problems. One difficulty in connection with the transfer of the Circular Test from machine tools to industrial robots is the analysis of the entire deviation. In the case of a cartesian robot or a numerically controlled machine tool only two linear axes are necessary to realize a planar circular movement. Industrial robots of most spread configuration have six independent axes to achieve the programmed position, therefore the question in which way single axes influence the test results cannot be answered easily. By fingering a plane surface, which is oriented perpenticular to the rotary axis, the orientation path accuracy can be calculated. By conscious variation of control circuit parameters influences of single rotary joints on the real path can be brought to light.

3. Path Characteristics

Basically path accuracy and repeatability definitions are independent of the command path shape. Path accuracy characterizes the ability of a robot to move its mechanical interface along the command path in the same direction n times, and also n times in the opposite direction (ISO 9283, 1988). The path accuracy is the maximum path deviation obtained in positioning and orientation. In this connection it is important to mention the main difference between path accuracy and circularity. The first property describes the capability of positioning during movement. Circularity means the maximum radial range Δr determined between the real path and the least square circle as a result of measured tracer displacements. As path accuracy depends on the fact that the measurement device centre exactly matches with the defined position, the centering operation must be carried out exactly. In the case of Circular Test the shape accuracy is of main interest, the eccentricity is calculated and can be directly corrected.

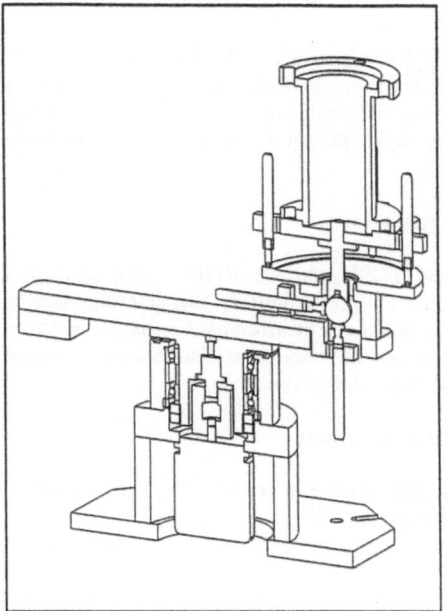

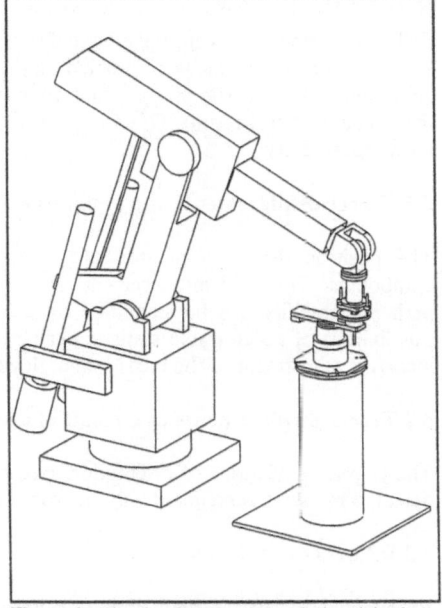

Figure 1: Tracer arrangement to detect deviations in different directions

Figure 2: Circular Test applied to industrial robots

4. Test Procedure

The measurement device is fixed on a stand situated within the work space and adjusted to the reference coordinate system of the industrial robot (Fig. 2). This can be achieved by the help of a probe connected with the mechanical interface, which is lead along the measuring arm. Next the centering operation must be carried out. One possibility is to position the ball, which represents the mechanical interface, to a cylinder surface at three points. The coordinates of the measurement device centre can be calculated as a result of the three locations. Following it is possible to correct the programmed positions according to the resulting eccentricity after the first Circular Test has been carried out. The robot programme to achieve the circular movement consists of two semi-circles defined by three points (start point, end location and one intermediate position). Two circles are programmed to ensure the detection of the rotary transducer reference signal. The test procedure at one position encloses measurements both clockwise and counter-clockwise. The fourier-analysis is an important aid to find analytic relations between velocity and the tracer results. The whole deviation between the actual three dimensional curve representing the ball movement and the nominal circular path can be divided into two parts. The translation at the mechanical interface is detected by two tracers, one in radial direction and the other arranged parallel to the rotary axis. The second part concerns the deviation in orientation as a rotation about the ball centre.

5. Test Results

Following some significant results are discussed, the test object was the industrial robot RIKO 106. The examinations were carried out after a warm-up period under normal operating conditions. A very significant point to evaluate the efficiency of a testing method is the repeatability of test results. Of course it is necessary to distinguish between short-term or long-term repeatability.

5.1 Tracer displacements as a result of position deviations

The position deviation of the tool centre point (TCP) consists of the radial and axial component. The axial measurements are represented as linear figures. The determination of path repeatability can be simplified by using the roundness measurement software with its possibilities of comfortable statistic evaluation. The first harmonic that also includes parts of incorrect adjustment of the measurment device, has been filtered out.

5.2 Tracer displacements as a result of orientation deviations

The angular deviations of orientation axes can be calculated as a result of the displacements, detected by two tracers fixed with the robot orientation part.

5.3 Influence of velocity

By examination of the deviations at different speeds it is possible to find out the maximum velocity, which allows to stay inside a defined accuracy range. In addition fourier-analysis enables a simple evaluation of path characteristics (Fig. 3).

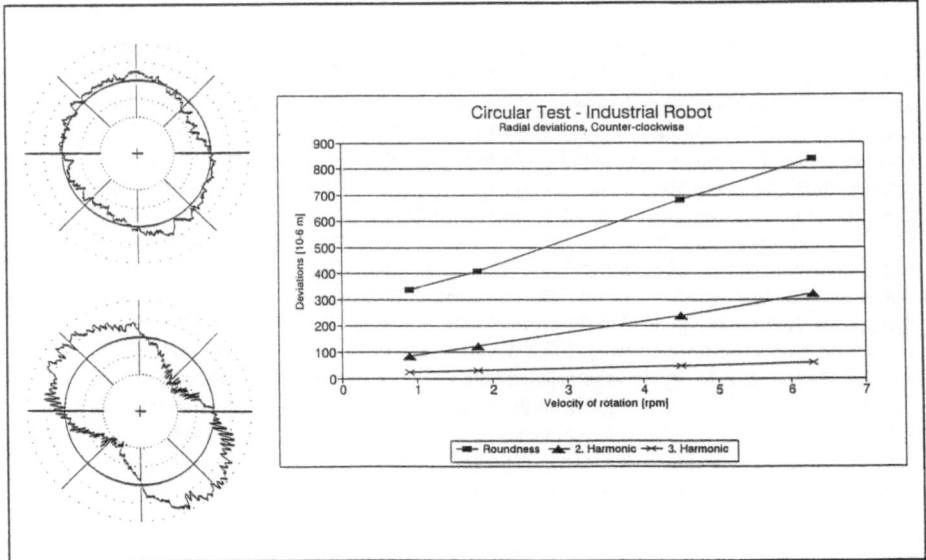

Figure 3: Influence of velocity by using results of fourier-analysis

6. Mechanism Design and Animation

The use of a powerful CAD-system to develop and optimize the measurement system offers some advantages. Beside the easy and rapid modifying and assembling of the construction, mechanism design is a valuable aid to evaluate and demonstrate test results. The task enables to create a mechanism by defining joints with specified degrees of freedom arranged between the according objects. Functions, which are the approximation of tracer displacements by using the results of the fourier-analysis, have to be related to the according translational joints to animate the system-movement (Fig. 4).

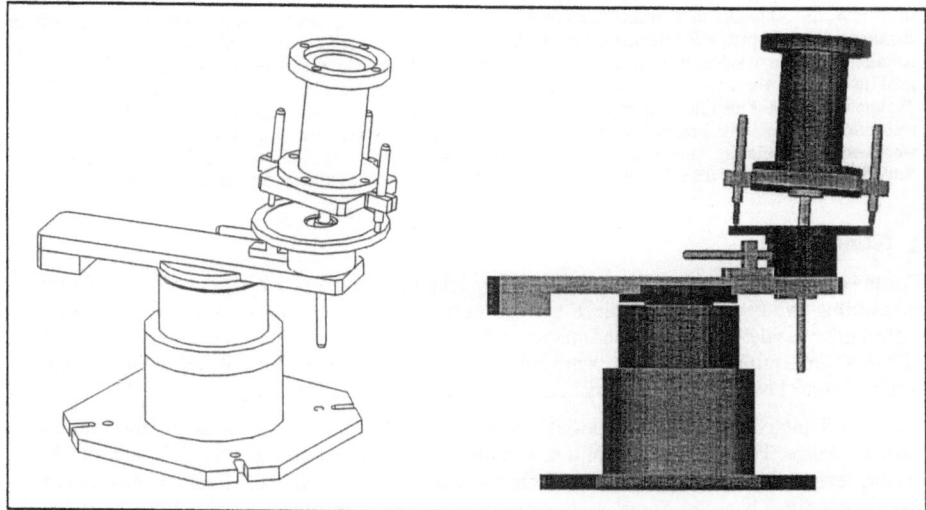

Figure 4: Views of a configuration using CAD-mechanism design

The described measuring technique represents a simple and universal method to characterize robot performance as well as machine tools. The additional advantage in testing machining centres is that the tracers fixed in the orientation part can detect geometric deviations caused by guides. Therefore it is possible to determine the guide shapes by using fourier-analysis in addition to radial deviations. A small radius (e.g. 30mm) is used to examine the quality of the control and drive system, other radial dimensions according the maximum axis movement (e.g. 100mm or 150mm) are suitable to determine geometric deviations. Now it is intended to carry out Circular Tests with industrial robots of different types and manufacturers and to adapt roundness software to special requirements of robot testing.

References

Knapp, W. (1992). 1. Grazer Präzisionstag. Der Kreisformtest und die neuen Abnahme-normen, 1-6

Knapp, W. (1986). Der Kreisformtest zur Prüfung von NC-Werkzeugmaschinen. Grundlagen, 1-18

ISO 9283 (1988). Manipulating industrial robots - Performance criteria and related test methods. Path characteristics, 29-33

A low cost robot system for stylus and workpiece manipulation in computer aided quality control

P.H. Osanna, E. Sarigel, N.M. Durakbasa, R. Oberländer, C.P. Heiss, D. Prostrednik
Vienna University of Technology

Abstract: The past twenty years have seen a continued increase in importance of computer aided measurement techniques as a means to control modern production engineering. Industrial robots are used for manipulating workpieces for inspection in CMM surroundings especially when high flexibility and low cost solutions are demanded as it is typical for small industrial plants. The prototype of a measuring cell with the goal to achieve the fexible automation of measuring tasks are drawn up at the Vienna University of Technology. A local area network of personal computers is linked to a small high precision CMM and to other measuring devices. The measuring cell also includes new developed and constructed handling systems for workpiece manipulating and automatic stylus exchanging. There are links to data banks for production, construction and quality data. The development of this system and the experiences gained are described.

1. Context

For the complete determination of a complex workpiece geometry are often a lot of different measuring points of the workpiece surface neccessary, which can only be reached with different stylus devices. For an automatic exchange of stylii, co-ordinate measuring machines (CMMs) are equiped with stylus exchange devices. These are an important step to an efficient quality control by more dimensional co-ordinate measuring techniques.

These exchange devices render it possible to measure different parts automatically by simultaneous ordance. There is no need for a new calibration and the exchange process is very quick. At the design process of such a system there are two goals to be followed. First the undercut of the market given boarders of the production costs. Second to share the resource between stylus magazine, stylus exchanger and the control module. The optimal solution satisfies the overall goal of maximal productivity.

Further the workspace of the CMM should not be reduced by the exchange device and the magazine should be sufficient covered.

Following different well known possible concepts of stylus exchange devices, which are positioned out of the measuring volume are compaired and a solution is presented.

2. Concepts of automatical stylus exchange systems

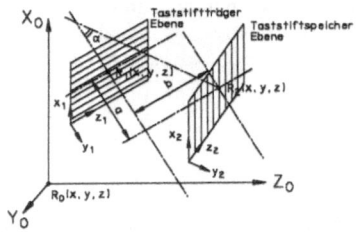

Figure 1: Arrangement of stylus storage, principle

Automatical stylus exchange systems are in order to serve a stylus or stylus combination at the right time, position and order to the CMM. Therefore the stylii have to be transferred on a closed track between a reference point of the CMM and the magazine.

The definition of the join direction of the stylus at the CMM and the magazine as vectors is neccessary for the determination of the adaptor (stylus holder) and the stylusstorage normal plane, which are decided for the kinematics of the handling action.

The totality of possible positions is formal given by the systematic combination of two vectors. For the qualification of the possibilities a minimazation of movements or degrees

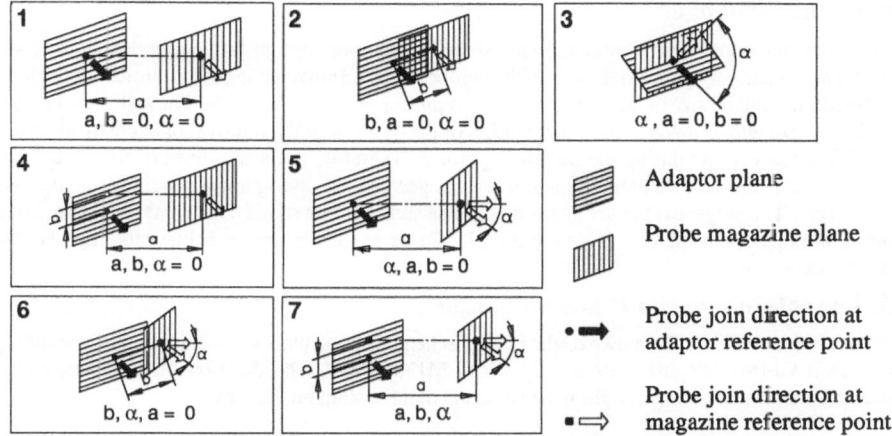

a: the distance between reference points, perpendicular to probe adaptor plane
b: the distance between reference points, in adaptor plane
α: Angle between join directions

Figure 2: Solution of position assignment of adaptor and probe storage

of freedom are used. That is the reason, why the join directions should be parallel or at least connectable with just a single turn.

For the describtion of a solution there are following parameters in use:

- the distance between the reference points, measured in the adaptor plane,
- the distance between the reference points, measured normal to the adaptor plane,
- the angle between the two planes.

The combination of these three parameters gives seven possible solutions. Of these seven possibilities one is impossible and the two, with the magazine on the table of the CMM, are standard solutions.

3. Stylus magazine

The different kinds of stylus magazines are avaliable. The distunguishing feature of these types is the capacity of the magazines to hold stylii. The magazine structure can be groupped in to the basic form of: Board, bar, disk, chain. The bar magazines may own a translatory preselected movement, which needs a corresponding space. The common feature of disk and chain magazines is a closed kinematic path for stylus preselection and therefore a preperation of the styluss for the exchange on a reference point. Both type have main differences in their constructive structure. The main feature of the disk magazine is a unextendable amount of storage places. On the other hand a chain magazine - like a CNC machine center - is constructed of single storage elements, which are directed in a path consisting of arcs and straight lines. Therefore disk magazins are characterized of the rotational preselected movement. For chain magazines such a movement has to be realized by an additional equipment. Very important for the choice of a certain magazine is the needed amount of the stylus places and the needed space for

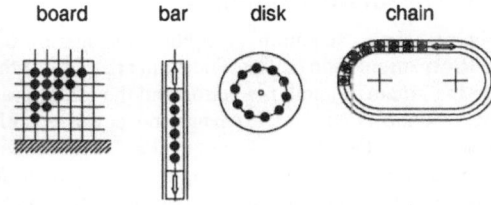

Figure 3: Various magazine structures

establishing the magazine. Basicly it is possible to distinguish between star, crown and trommel stylus combinations and between horizantal and vertical join movement.

210

4. Stylus exchanger

The exchange equipment of an automatic stylus exchanging system influences deeply the side time.The exchanging time is dependig of the number of asked movements, or the number of degrees of freedom , and the size of the path and the angle during changing the stylus. The stylus exchanger is an activ system, it means, that a preparation of the new one and a removal of exchanged stylus configuration is done during the measuring proces. Therefore put aside times of the CMM are reduced to a minimal value. The stylus exchanger can be configurated as single or double exchanger. The single exchanger handles only one stylus or one stylus combination. This is from the economical point of view not very useful. Shorter exchange time is achieveable by using a double exchanger.

5. The solution variants and valuation

In this chapter there will be shown on the basis of schematic principles the basic proposals of solution for small CMMs /Neumann,1988 - ÖNORM M1386,1900/. The decision of possible system realisation in chapter 6 is given from the valuation of these solution variants.

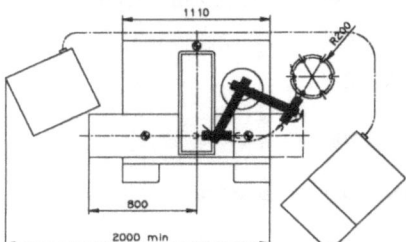

Figure 4:   rotation-rotation-link (RR)
manipulator - (1)

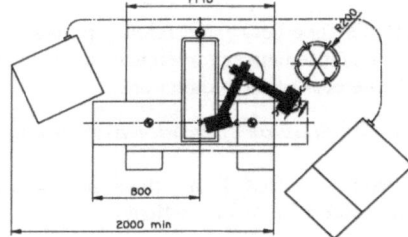

Figure 5:   translation-rotation-rotation-trans
lation-link (TRRT) manipulator - (2)

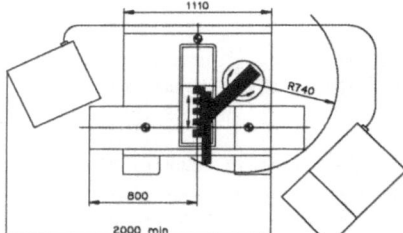

Figure 6:   rotation-translation-link (RT)
manipulator - (3)

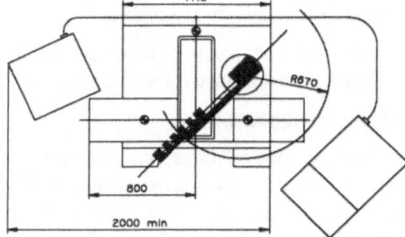

Figure 7:   translation-rotation-translation-
link (TRT) manipulator - (4)

5.1. The solution variant 1

Figure 4 shows an automatic stylus exchanger for 6 stylus combinations, which is developed as RR manipulation device. The main problems of this solution, with two degrees of freedom, are the kinematic path mediation and the grasping a stylus from the magazine. The premise of this solution is, that the exchange movement is horizontal. This causes the unique kinematic movement of the adapter. Therefore the location of the magazine is determined. Another problem appears when grasping or storing a stylus combination from/to the magazine. Because of the missing vertical movement (two degrees of freedom) and unexpected deviations of the circular movement the exchanging process can be disturbed by friction.

5.2. The solution variant 2

The automatic stylus exchanger for 6 stylus combinations according figure 5 is developed as TRRT manipulation device is an extension of the first solution variant. Grinding and with its possible damages of the magazine or adapter will be avoided by the integration of the vertical movement. This resultes in simple construction of the disk magazine and the storage can be placed at any position.

5.3. The solution variant 3

The solution 3 shown in figure 6 provides an automatic stylus exchanger for 4 stylus combinations. The design is a RT manipulation device. In this solution, which works with only two degrees of freedom, the stylus magazine is integrated in the rotating device.

During the measuring process the arm is moved out of the measuring range in a waiting position. The changing process starts by rotating the arm and caused by the circular movement the geometrical relations of the magazine places are exactly defined. Therefore the location of the exchanger is also determined.

Because of the great rotating radius, the limited amount of magazine places and the problem of the rotating under the adaptor mounted on the stylus head (sliding, friction) this solution is not the best one.

5.4. The solution variant 4

By the extension of one more degree of freedom in vertical direction (translation link) and by changing the arrangement of the magazine places relative to the rotating and translation direction (figure 7) the automatic stylus exchanger as TRT manipulation device with 5 to 6 magazine places solves the problems with advantageous geometrical relations. Also in this solution the location of the exchanger is determined clearly because of the unchangeable geometrical relations if they are once selected. The positive aspects of this solution are the simple control of the system and the suppression of a transmission unit for the stylus between measuring machine and magazine. Also the limitation of the measuring range can be reduced by selecting a propriate location; but the limitation is greater than in the former solutions. It is possible to increase the amount of magazine places by displacing the location of the exchanging unit.

6. The realisation of the stylus exchanging unit

At the department "Austauschbau und Meßtechnik" Vienna University of Technology a prototype, according to solution 2, was developed to increase the flexibility and to improve the adaptability of a CMM with column construction. Figure 8 shows the layout of the total system.

The complete automatic stylus exchanging unit consists of:
- stylus magazine,
- double gripping arm stylus exchanger,
- PC to control the step motor and the connection with the CMM computer,
- control unit for the motion of the exchanger and the magazine.

The circular stylus magazine has 6 storing places with a stylus distance of 160 mm, the resulting diameter of the storing disk is therefore 320 mm. The motion of the magazine is executed by step motor and gear unit.

The TRR manipulation device is a stylus exchanger, which is equipped with a double gripper, with 3 degrees of freedom. Each axis is driven by a step motor.

The computer has two tasks:
- Control and inspection of stylus exchange system.
- Communication with CMM-Computer by the serial RS-232C interface.

212

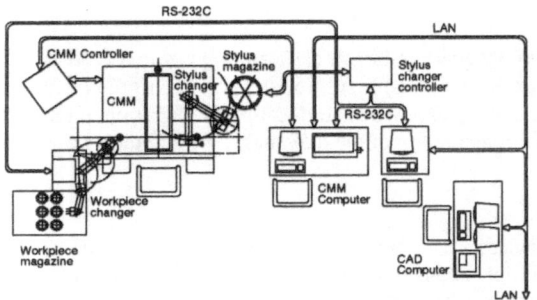

Figure 8: CMM with stylus and workpiece exchanging units

The exchange command of the CMM-Computer to the control computer releases the following operating sequence:

- Rotating of the magazine disk with the demanded stylus in a predefined exchange position.
- Assuming the stylus by the exchanger.
- Interruption of the measuring process, moving of the stylus head to its exchange position by simultaneous movement of the table out of the rotating range.
- Rotating the exchanger in direction of the stylus head by simultaneous rotation of the gripping unit.
- Assuming the stylus combination, which should be changed, by the free gripper, disconnecting the adaptor from the stylus head.
- Rotating the gripping unit 180 degrees, connecting the new adaptor to the stylus head.
- Rotating the exchanger out of the measuring range, starting the further measuring process.
- Turning the gripping unit, storing the stylus combination in the provided place of the magazine.
- Exchanger goes into waiting position.

The control unit of the stylus exchanging system is connected with the control computer by serial interface (RS-232C).

7. Concluding remarks

The integration of a CMM in flexible manufacturing systems is a good deal by using an automatic stylus exchanger. The stylus exchanging unit in combination with a proper stylus magazine provides the possibility to measure different workpieces in any sequence without interruption. This allows the increased use of CMMs in manufacturing zones resp. in shifts with poor or without operation service /Eiden,1988/.

Because of the valuation of the solution variants and according to the demand of a fixed location of the stylus head during the stylus exchanging process and according to the restriction to a certain CMM mainly the solution variants 2 (figure 5) and 4 (figure 7) are possible candidates for establishing at low costs. The adation to other types of CMM construction, especially for the gentry type, is possible.

Bibliography

1   Eiden,H.P.: Einbindung der Koordinatenmesstechnik in die flexible Fertigung, Erfahrungsberichte und Konsequenzen aus der Sicht eines Systemanbieters. VDI-Bericht No. 711: Fertigungsmesstechnik und Qialitätssicherung, S93/111, 1988

2   Hahn,H.: Wichtige Komponenten zum flexiblen Messen auf Koordinatenmessgeräten.wt Werkstattstechnik - Zeitschrift für industrielle Fertigung 74 (1984), Nr.9, pp.539/544

3   Neumann,H.J.,Trittler,P.: Fertigungsnaher Einsatz von CNC- Koordinaten-Messgeräten. Werkstatt und Betrieb 117 (1984),No. 9), pp. 573/578

4   Neumann,H.J.: CNC - Koordinatenmesstechnik. Expert Verlag, Band 172, 1988

5   ISO/DIS 10360/2: Data Sheet for Coordinate Measuring Machines. ISO/TC3 N530, 1992

6   ÖNORM M1386: Koordinatenmeßtechnik; Geometrische Grundlagen; Benennungen und Definitionen zur Werkstückgeometrie

Low Cost Automated Measuring System
For Circularity Measurements

B. Acko, M. Milfelner, A. Sostar
Laboratory for Industrial Measurements
Department for Mechanical Engineering
Faculty of Technical Sciences
University of Maribor

Abstract: The measuring system that is being developed in our laboratory was designed to solve a special problem of circularity measurements. Since the workpieces to be measured (rollers of great dimensions for paper and similar materials) cannot be fixed in the center points in the axial direction, a modified principle has been developed and is still being tested. The roller lies on two supports and the deviations are measured by one or two probes (two different principles). The principles of the device design and of the calculations are described in the paper.

1. Introduction

The device that is being developed in our laboratory for a certain customer fulfils all requirements of the modern mesauring technique. It will be equipped with a microcomputer for the data calculation and statistical process control (SPC). All the geometric aspects (dimensions, shapes, tolerances, fixtures etc.) and the environment conditions were carefully examined and studied. The corresponding measuring standards were studied as well. The device is designed to be used directly on the machine tool.

2 The Measuring Problem

2.1 Standard Methods for Circularity Measurements

If it is possible, such methods are chosen for the circularity measurements that correspond to the circularity tolerance definition. There are three main types of such methods:
- measurements on special devices for circularity
- measurements on three - coordinate measuring machines
- measurements of radial deviations on the workpiece that is fixed in center points (in axial direction)

2.2 Limitations in Circularity Measurements

The standard measuring methods cannot be used on all workpieces. Especially workpieces of great dimensions cause a lot of troubles. They can neither be put on a coordinate measuring machine nor on a special circularity measuring device. In many cases such workpieces can even not be fixed in the center points because of their dimensions. The weight of the workpiece and the machining conditions can also be a good reason for not fixing it in its center points.

In the above cases a modified measuring method should be used. The only possible solution is to support the workpiece in the measuring cross section by one or more supports and to measure the deviations in relation to the supports. In fact, many problems can appear in this method. The deviations shown by the measuring probe are not real circularity deviations. In some cases no deviations are shown on workpieces with great circularity errors. Two examples are shown in Figure 1. A two-point measurement of an oval workpiece is correct (the shown result should be divided by 2), but a three point measurement ("v" support) gives a completely wrong result (the shown result is 0 although the workpiece is not round). An opposite example can be seen on the triangular workpiece (Figure 1).

3. Measuring Problem to be Solved

We should find a proper solution for diameter and circularity measurements on rollers (for paper rubber, etc.) of great dimensions (several meters). The rollers cannot be fixed in the center point and the measurements are to be executed on the machine tool. The measuring results should be calculated by a computer, and a statistical process control should be provided as well. Circularity is to be measured in many cross sections and the tolerances are very small (about 0.01 mm).

4. Measuring Device and Data Evaluation

4.1 Draft Design

The design of the measuring device without numerical parameters is shown in Figure 2. The first solution, which is shown on the left side, is the following: the measured workpiece rotates on two supports and deviations are measured by an inductive (or laser) probe. The other solution is similar. A measuring frame with probes lies on the workpiece which is rotating. The deviations are measured by two probes (an additional probe is used to detect different kinds of shapes such as oval, triangular, five angular etc. shape). This device is handier and designed to measure different diameters (Figure 2).

The angle between the supports and the angle between the support and the probe will be calculated from the simulation results. The angles have not yet been precisely defined.

4.2 Data Evaluation

4.2.1 Definition of the Division Angle

A division angle is a workpiece rotation angle between two measuring points. It depends on the number of the measuring points. The following values gave the best results in our simulation: 15° (24 measuring points), 5° (72 measuring points), 3° (120 measuring points), 1° (360 measuring points).

4.2.2 Definition of the Support and the Probe Angles

The angle between the supports must be a multiplier of the division angle, and the least denominator with the value 360° should be as great as possible. The angle between the probe and one of the supports must fulfil the same requirements as the angle between the supports (Figure 3)

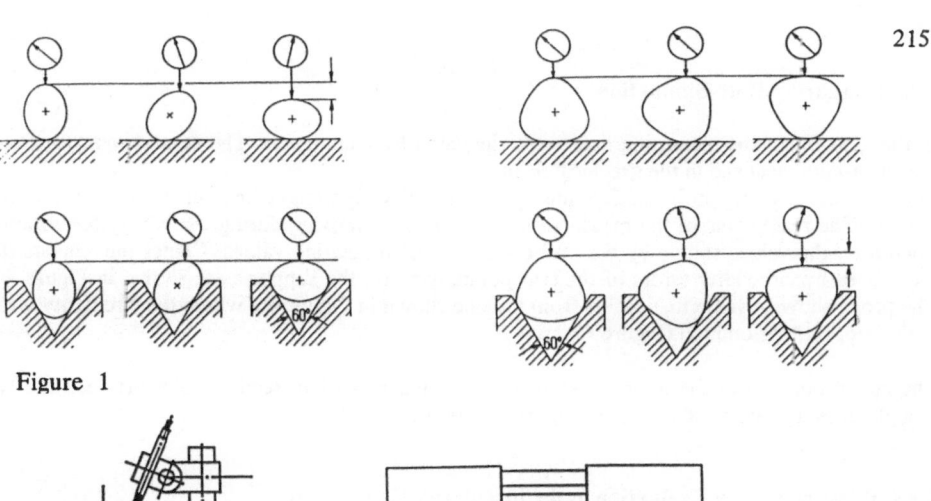

Figure 1

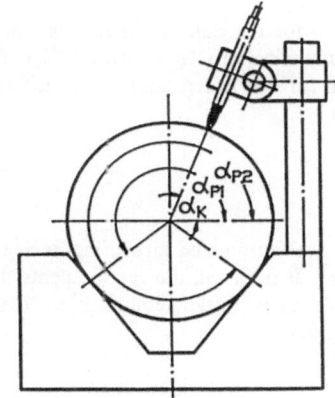

Figure 2

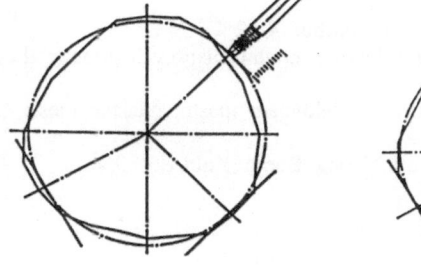

Figure 3

Figure 4

4.2.3 Measuring Data Simulation

In the experiment the measuring data were simulated by a computer. The results were printed in the alpha-numerical and in the graphical form.
The supports and the probe surface are approximated with straight lines in the measuring cross section. The real center of the measured circular shape is moving during the workpiece rotation Therefore, the values shown by the probe are not real measuring values. Center movements that are caused by circularity errors of the two points lying on the supports are shown in Figure 4 b The probe shows a value that differs from the one shown in Figure 4 a where the circularity error in the supports are omitted (Figure 4).

The actual position of the moving center is calculated as an intersection point between the two straight lines that are parallel to the support directions.

4.2.4 Measuring Data Evaluation and Ciurcularity Calculation

The measuring error depends on the division angle, the angle between the supports, and on the probe angle (see Figure 3). The error is of the first degree, and the values lie between 0% and 300% of the shown value. The system of n linear equations with n unknown values is used for the error correction. The number n is equal to the number of measuring points.
The center and the diameters of the tolerance circles are calculated from the corrected measuring values using the least square method.

5 Process Automation

The measuring process is completely automated. The software for the data evaluation is almos finished. A commercial low cost software will be used for the statistical process control (SPC). The microcomputer, which will serve for the calculations and for the integration into differen information systems and data bases, is still being developed.

6. Conclusions

The measuring system described in this paper represents a low cost automated measuring device o high accuracy, which can be used in different fields of industry. It fulfils all the requirements tha are set to modern industrial measuring devices and represents an important contribution to the development of the Slovenian industry of automated measuring devices.

7 Bibliography

Warnecke, H.J., Dutsche W. (1984). Fertigungsmesstechnik. Springer-Verlag Berlin, Heidelberg 117-121, 361-367
Sostar, A.(1992). Tehnolo{ke meritve. Zapiski predavanj, Maribor, 4-42
Pogorevc, B.(1992).Merjenje in digitaliziranje na koordinatnih merilnih napravah. Master thesis TF Maribor
Acko, B. Natancnost trikoordinatnih merilnih naprav na delovnem mestu. Master thesis, Tl Maribor
DIN Taschenbuch 11.(1991). Längenprüftecnik 1. Beuth Verlag, Berlin, Köln, 330-354

Rationalization of the CAD Data Management in Quality Assurance Systems

B. Sesko, A. Sostar, A. Acko

Laboratory for Industrial Measurements
Department of Mechanical Engineering
Faculty of Technical Sciences
University of Maribor

Abstract: Since very complex workpieces are measured on the coordinate measuring machines (CMM), a lot of nominal data are needed and a lot of measuring data are produced. Measurements of workpieces with sculptured surfaces represent a specific problem.. Therefore, an efficient data transfer system between the CAD system, the CMM and the manufacturing process is needed. An example of a very good interface for the mentioned data transfer is the HOLOS system (a product of the German company Zeiss),which is being studied in our laboratory. Some of the results of our research in the field data transfer are presented in the paper.

1. Introduction

Sculptured surfaces represent a very special and important problem in the dimensional measuring technique. Their mathematical description is very rarely known. For this reason, different mathematical methods based on the coordinates of few characteristic points for the description of the sculptured surfaces are being developed. The development of programme systems for digitizing (calculation of mathematical models of the sculptured surfaces) is very intensive as well.
In the paper some results of the practical and theoretical researches in the Laboratory for Industrial Measurements are presented.

2. Data Connection Between the CAD Systems and the Coordinate Measuring Machine (CMM)

In each stage of a manufacturing process (design, manufacturing planning, manufacturing, quality control) the data from the preceding and the following stages are used. Therefore, a certain amount of data always circulates.
If all the stages are aided by computers, the data connection between those computers is necessary. Every computer in the network should have access to all the needed data, although some of the data were produced and stored on other systems. If the data connection is adequate, a significant control rationalization is achieved because unnecessary data duplication is avoided (Figure 1).

3. Sculptured surfaces - New Requirements to the CMM

The first step in the formation of a sculptured surface is usually made by proceedings (design, modelling. ..) where the result is not a mathematical description but a real model of a product.

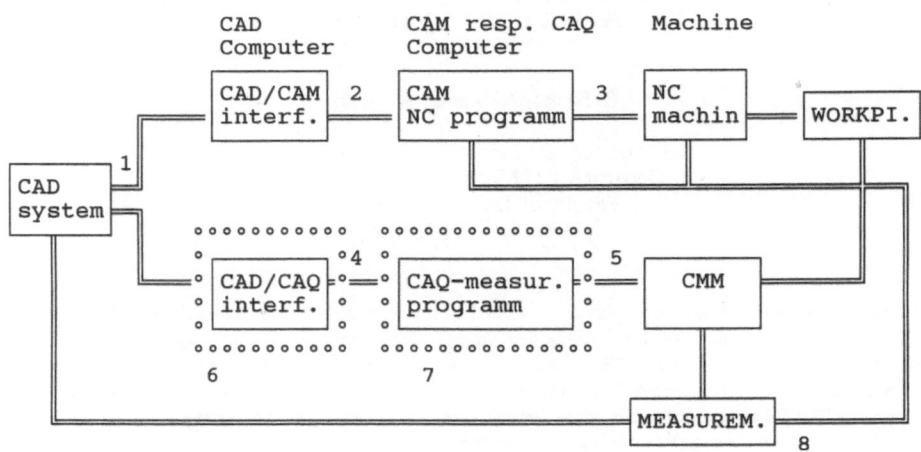

Figure 1

Legend:
1 - geometry
2 - surface in CAM format
3 - programme
4 - surface in CAQ format

5 - programme
6 - surface in CAQ format
7 - on-line or off-line transfer
8 - deviation determination

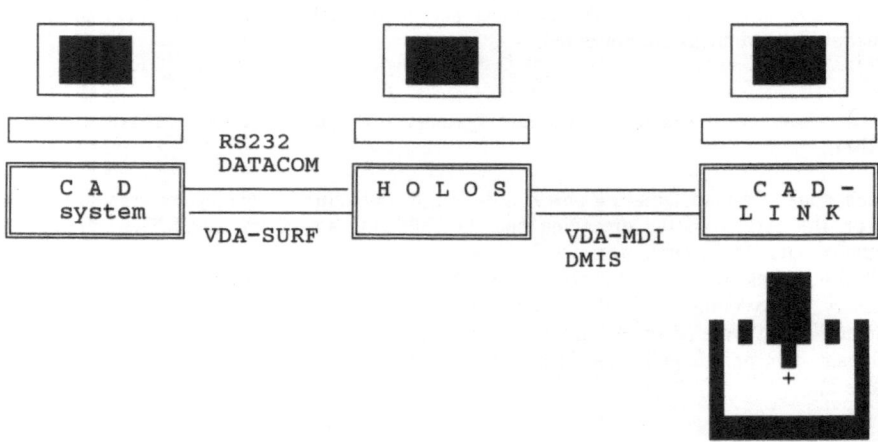

Figure 2

The standard measuring software for measurements of regular geometrical elements (point, straight line, circle, sphere, cylinder etc.) is not sufficient any more. Programming systems for measurements and digitizing of sculptured surfaces base on complex mathematical methods (Bezier, B-Splines, Coons...). The mentioned methods are usually based on surface splitting into smaller subsurfaces, which can be described by much simpler mathematical formulas.

An example of an advanced system for measurements and digitizing of sculptured surfaces, which can be linked to different CAD systems, is the HOLOS system produced by the German company Zeiss. This system has been ordered by our laboratory and is expected to arrive very soon. It will be used for research and development purposes. The goal of our research is CAD/CAM/CAQ data connection development. The data of the CAD system, the CMM, the HOLOS system and the flexible machining centers will be connected.

HOLOS was developed to improve the measuring and digitizing possibilities in the industries that produce products with sculptured surfaces (elements that are exposed to the fluid and air flows, ergonomically designed products, aesthetic products, products with special functions, etc.).

HOLOS is able to communicate with all CAD systems that generate and receive data in the VDA-FS format (Figure 2).

The geometrical data are transferred in the VDA-FS format and the technical instructions in the DMIS format.

The conversion of the VDA-FS (SURF) format to the VDA-FS (MDI) and vice versa is based on the Bezier method in which the surface is split to small subsurfaces - patches. Each patch comprises a certain number of measuring points.

The basic purpose of the HOLOS is to be an interface between the CAD system and the CMM, or in other words, between the mathematical and physical model.

4. Conclusion

As it is seen in this paper, the main trend of the development in the CMT and the QA is the connection between the CAD, CAM, and the CAQ systems. This trend is followed in our laboratory where the complex system of data connected subsystems CAD-CMM, CAD-CAM-FMS, and CMM-FMS is being developed. The basic expected goals of the mentioned development are the following: reduction of the circulating data, better data accessibility, required product quality assurance and maintenance, scrap reduction by means of the statistical process control and corrections of NC programmes and machine tools, quality control by the minor closed loop (CMM-machining center) and the major closed loop which includes the CAD system

Bibliography

Bruno, S., Grebe, W.(1990). Freiformflächen graphisch - interaktiv digitalisieren. Sonderteil in Hanser Fachzeitschriften - CAD/CAM/CIM. Hanser Verlag, München, 146-154

Garbrecht, Th., Roth, S.(1989). Strategien für das Messen von Freiformflächen. VDI Bericht 751. VDI Verlag, Düsseldorf, 303 - 319

Georgi, H.(1992). HOLOS - Messen und Digitalisieren von Freiformflächen in der Automobilindustrie. Workshop Ständermesstechnik, Riegelsberg

220

Heeg, H.(1992). Anbindung von Koordinatenmessgeräten an CAD Systeme. Workshop Ständermesstechnik, Riegelsberg

Pogorevc, B.(1993). Merjenje in digitaliziranje na koordinatnih merilnih napravah. Master thesis, TF Maribor

Sigle, W.(1990). Messen von Freiformflächen-praktische Einsatzerfahrungen. VDI Berichte 836. VDI Verlag, Düsseldorf, 19 - 32

Sostar, A., Pogorevc, B.(1993. Einsatz von Koordinatenmessgeräten im CAD/CAM/CAQ - Verbund. Technik Report, Sonderteil Fertigungstechnik, 3/93

Weule, H., Klein,H.(1990). Messen und Digitalisieren von Freiformflächen unter Einsatz von CAD Systemen. VDI Berichte 836. VDI Verlag, Düsseldorf,. 1-17

Acko, B. (1989). Natancnost trikoordinatnih merilnih naprav na delovnem mestu. Master thesis, TF Maribor

Modular Parallel Gripper System

G. Kronreif, R. Probst
Institute for Handling Devices and Robotics
Technical University, Vienna

Abstract: A new, programmable and flexible gripper system with position-, force- and speed-controlled grip jaws has been developed by the Institute for Handling Devices and Robotics of the Technical University of Vienna in co-operation with the division of System Engineering and Automation of the Scientific Academy of Lower Austria as well as the Robofil Company in Wels, Upper Austria.

1. Introduction:

In spite of the current economic situation and the call for 'lean production' in European companies, the next jears will bring an increased demand of automated, roboterized assembling systems. The required flexibility of these assembling cells will be necessary for the gripper systems, too. In most cases, the optimal design of the utilized systems is responsible for productivity and effiency of the assembling system.

In an opinion poll of the scientific magazine "ROBOTER" (1992), the cell element 'gripper' has become very good marks. Gripper systems seem to be full-blown components. To meet the expected requirements, flexible gripper systems have to be developed.

To take the first step, a new gripper system with position-, force- and speed-controlled grip jaws has been developed by the Institute for Handling Devices and Robotics of the Technical University of Vienna in co-operation with the division of System Engineering and Automation of the Scientific Academy of Lower Austria as well as the Robofil Company in Wels, Upper Austria.

2. Aims:

One of the most important parameters is the grip region. Conventional gripper systems (with a similar weigth) have a gripper stroke up to 15 [mm] per gripper jaw. If the dimensions of the gripped parts vary in an extended scope, a gripper change is indispensable - the assembling system is getting more costly (several gripper systems, gripper change device, etc.). Therefore, one of the main goals was a considerable increasing of the gripper stroke.

Another important parameter is the grip force. To reach a higher flexibility concerning the strucure of the part, it is necessary to vary the grip force for each gripping process.

Traditional gripper system are picking up the parts centred. If the parts vary in measure, the part will be moved during gripping process. This movement often leads to a damaging of the feeding of parts (e.g. magazine). Therefore, SERVO-GRIP should take the parts aligned to a defined (i.e. teached) reference plane.

3. Construction:

Each module (i.e. each grip finger) has a separate, position-, force- and speed-controlled axis. The grip finger is moving along a linear unit and is driven by a small DC-motor and a no-backlash ball screw. The careful selection of the mechanical components and their compact arrangement are leading to a sufficient stiffness of the mechanical construction, and to a suitable gripper dimension and mass.

Each single grip process can be 'teached in' by means of the gripper control system - an IBM XT/AT compatible personal computer with an extension board for motion control. Furthermore, the PC serves as a storage medium for the programmed grip processes (see Fig. 1). The software of the control system for the teaching- and the operation-phase has an user-oriented graphical, window- and menu-driven interface. Each feature of the gripper system (open or close the gripper, center the gripped piece, etc.) can be called by a supervisory control system or by the robot control (i.e. by means of some of the digital I/Os of the robot control).

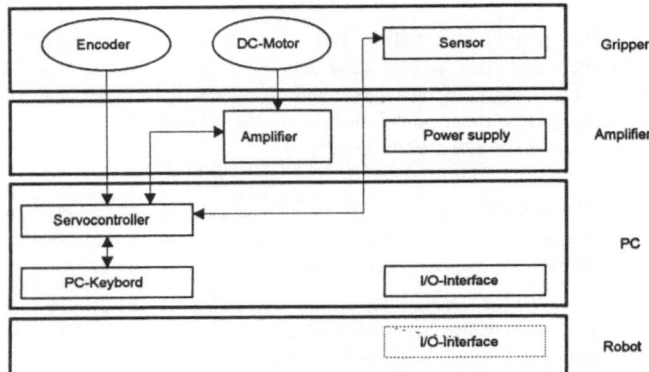

Fig. 1: Control system of the gripper system

The gripper system SERVO-GRIP should be utilized by small and medium sized companies, too. Therefore, we have not used (expansive) sensors.

4. Operation:

The parameters of each grip process can be set during the teaching phase. First the dimensions of the piece (maximum and minimum size), the necessary grip force, the gripper play (distance between the grip finger and the piece for the open gripper), and the gripper process speed of each of the single grip processes should be adjusted. Furthermore it can be determined whether (see Fig. 2):

⇨   to center the piece after gripping

⇨   to move the gripped piece within the gripper

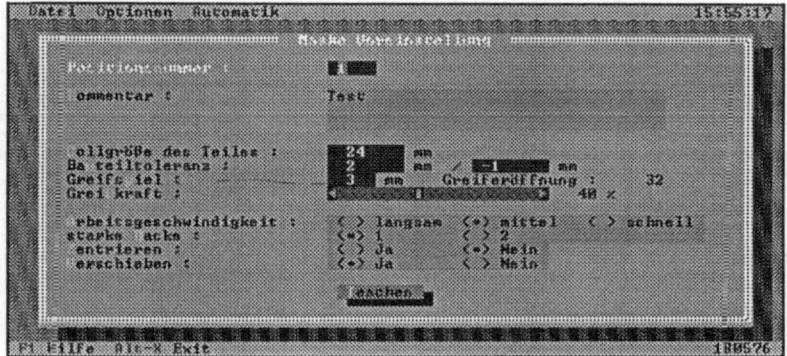

Fig. 2: Input of the gripping process parameters

After setting the parameters mentioned above, the proper teaching of the grip process has to be accomplished - the grip fingers are moved into their grip position.

During the whole teaching phase the most important data, like the current distance between the grip fingers, the difference between the current, the calculated and the geometrical grip centre, etc. are displayed in order to support the user.

Furthermore, several functions are available. For example there is a test-button, which initiates a 'simulation' of the grip process with the parameters set before.

5. Applications:

➲ Centred - Aligned to a reference plane:

As mentioned above, the mechanical separation of the single grip jaws makes it possible to move the gripped piece within the gripper. If the piece has a defined reference plane and the piece has to be positioned referring to this plane, this feature is very important. Conventional parallel grippers can take the parts only centred. So the variation in size of the gripped part is corresponding to the position error of the robot.

Our gripper system SERVO-GRIP is picking-up the workpiece aligned to the reference plane and centring the piece **after** gripping (if needed).

➲ Gripper and measuring device:

The position of each grip finger is known any time with an accuracy of 5/100[mm]. So the gripper can be utilized as a measuring device. The dimension of the gripped piece can be compared with the allowed dimension tolerances (entered in the teaching mask). If the workpiece has not the correct size, the gripper control system can take appropriate measures - e.g. report the error to a quality control system.

➲ 'One for All' :

A further advance of this gripper system is its large grip region. It is possible to pick-up parts with a maximum dimension of 85[mm]. To grip parts with various dimensions within the minimum grip time (0.5[sec]), the grip play can be defined individually for each part.

Therefore and because of the possibility of an individual grip force adjustment, utilization of

224

this gripper system reduces the number of necessary gripper changes. The application will become more time- and cost-saving.

➲ **Force and 'sense':**

The gripper system SERVO-GRIP is able to pick-up the pieces precisely and shockless. Therefore very sensitive pieces can be gripped, too.

6. Technical Data SERVO-GRIP:

(Gripper with two movable jaws)

grip force	adjustable, up to 250 [N]
maximum grip region	type a: 60 [mm], type b: 85 [mm]
accuracy of measurement	5/100 [mm]
grip time	adjustable, minimum: 0,5 [sec]
main dimensions (WxHxD)	type a: 120x95x64 (without gripper finger) type b: 150x90x100 (without gripper finger)
weight	type a: 2 [kg]; type a: 2,5 [kg];

Tab. 1: Technical Data SERVO-GRIP

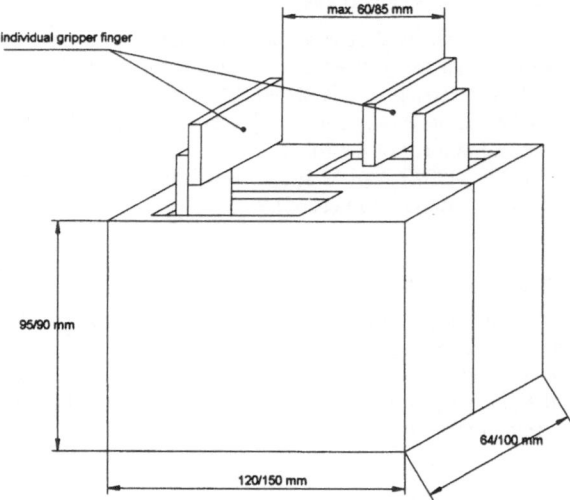

Fig. 3: Main dimensions (type a / type b)

The described gripper system is now available as a prototype. At the moment we are going ahead with several test series to determine the relevant characteristics of the gripper system (i.e. repetitive accuracy of gripper, accuracy of measurement, long-time test with steady load, characteristic line of grip power, etc.).

Example: Measurement of the positioning exactitude of the grip finger

The idea of this test was to proof the gripper mechanism and the electronic devices. The grip fingers were positioned (in teach mode) and the difference between real and displayed position was measured.
Figure 4 shows the maximum, the minimum and the average position errors (we made 20 measurements for each finger).

Result: maximum average error (absolute) : 0.02[mm]
            maximum average error (relative):   0.35 %

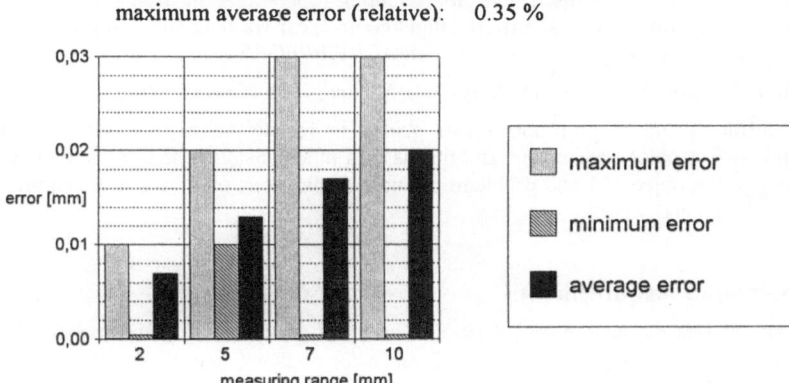

Fig. 4: Positioning exactitude of the grip finger

The experiences recieved from the first applications and the very positive response to the capacity of the gripper system indicates, that the developed gripper system SERVO-GRIP enables a further step towards an efficient use of industrial robots in assembling automation.

Robotics in Transportation

R. Genser

Austrian Federal Railways, Vienna, Austria

Abstract: Based on the results of the project "Austrian General Conception of Transportation", objectives for the use of robots in transportation systems are outlined. Application's areas and technical as well as psychological problems for the use of robotics in transportation are pointed out.

1. Introduction

The production of vehicles was one of the first fields for the use of robots. The manufacturing of cars has stimulated the development of robotics to a great extent. In the operational level of transportation, automation is used like for shunting operations in freight yards, track maintenance, luggage distribution at airports, flight control, car washing etc., but robots can be found scarcely as it was recognized also at the 5.IFAC/IFIP/IFORS

Symposium on Control in Transportation Systems, Vienna, 1986 (Genser,1986).

Meantime, further progress in robotics was gained in respect to expert systems, neural networks, microelectronics, sensors etc. But this has not much changed the use of robots up to now. Following, the objectives and problems for new applications of robots in transportation systems should be pointed out.

2. Objectives and Requirements

According the results of the project "Austrian General Conception of Transportation", see (Genser,1987), following objectives are of relevance for applying new technical solutions in transportation.
Because of limitations of human, of regulatory measures, or of constructional investments, transportation systems can be improved concerning efficiency, economical aspects, and social as well as environmental impact only by early use of new technical solutions. This is stated as one of the objectives of the Austrian policy for transportation.

Objectives in detail will be in respect to efficiency, efficacy, economy, social acceptance and satisfaction, limitation of environmental impact, and considering sustainable developments as well as according the distinctive features of the use of robots, like mobility:

- dependability, depending on quality
  robustness
  adaptability
  safety, meaning risk level is less than or equal to the specified tolerable risk
  integrity
  maintainability
  reliability

- high economy
- high efficiency
- high efficacy, also considering physical and mental effort
- ergonomic solution, also improving accessibility for disabled
- sustainable development considering the whole life-cycle from design to maintenance and scrapping.

The aspect of humane solutions is important, also concerning stability of solutions. But this aspect is two-sided. Damage, injury, or loss should not be caused by the robot of course, but the use of robots can improve these aspects for the transportation system itself, e.g. avoidance of theft of goods, loss of human live because of accidents at coupling of waggons is reduced etc.

The safety requirements may be more severe in comparison to the use of robots in a factory because of the mobility, the flexibility of use, and by the more open and less controllable environment. The robot has to fulfill, among others, the requirements:

- movement has to be in such a way in controllable and uncontrollable situations that in case of possible harmful situations free space is available for not doing harm to living beings otherwise the possibility to escape should be given by signals, time range, and design even for children, disabled, and untrained persons

- to cause no danger by contact or radiation, e.g. electricity, magnetism, light, heat, gas, cutting shape etc.

- no dangerous clutching, catching, cauhting, sucking etc.

If a mobilator can make dangerous moves without control of actuator given, a monitoring by a person or a monitoring centre has to be assigned.

The use of robots in the operational level of transportation requires resistance of parts of robot against high forces, acceleration, vibration, extreme weather conditions, and robustness against ice, snow, and dust.

The speed of movements, acceleration, and forces of robot itself will be in the range of the ability of human in general.

3. Impediments for the Use of Robots

Of course, innovation and technology transfer are much influenced by the cultural and societal conditions as well as affected by the engineering and production systems, taking in account the time needed for learning processes. Because of the international aspect of transportation a world wide interdependance is given. Road transportation has an extreme diversification of decision-makers, like authorities, car manufacturer, road construction industry, forwarding companies, private drivers etc. Organization and management are very complex. Railways would have more rigid organizations. But they are national oriented to some extent and the efficiency of international organizations for cooperation of railways is not very high. Moreover, railways are traditional organizations, which with only few exceptions are able for smooth adaption to new conditions. Air transportation has the best conditions given for innovation and adaption to new technologies.

Knowledge, awareness and skill are preconditions for successful innovations. Shortcomings of these requirements may be given in traditional organizations with limited competition or lack of consequences in case of wrong way of acting.

Human are reluctant to changes of well adapted practice especially if the positive effects of a new operation are not obvious, not recognized by own experience, or may result changes in other sensible areas like change of jobs or amount of votes.

This fact was evident in the case of the "Austrian General Conception of Transportation" when the measure proposed to investigate the use of robots for coupling freight cars. One of the objectives was to avoid the awful accidents at this job. This measure was supported by the Chamber of Labor, but it had to be withdrawn by demand of the Labor Union of the Austrian Railways.

4. Application Areas

According the objectives for transportation and the dissatisfaction given, robots may be suitable for improving the situation for following areas in the operational level of transportation:

- quick goods transfer in multimodal transport for distributed operations, but for central hubs, automatic systems would have higher efficiency

- handling of dangerous goods

- desinfection and sanitary operations

- mixed cargo or parcel handling

advantages would be:

> reducing loss by human-less operation
> increasing efficiency
> improving ergonomic aspects
> better quality
> higher flexibility in comparison to other solutions

- coupling of waggons, in Western Europe because the use of automatic coupling devices is blocked by political and

> financial conditions

advantage would be:

> reducing accidents
> improving efficiency
> stimulation of further development in robotics

- sewage handling for vacuum toilets

- cleaning of vehicles interior

advantage would be:

> ergonomic and sanitary improvement
> better quality
> higher efficiency
> higher flexibility

Robots would be usable like for catering in passenger service. But it has to be recognized the psychological aspect and the better economy of other possible solutions.

Robots would allow better flexibility and improve access to transportation if a solution can be developed for guiding and supporting disabled persons.

5. Approach

In accordance with the requirements given, generally a robot for the operational level of transportation will not only consist of sensors, information processing unit, decision making device, control units, actuators, and mobilators. Monitoring by a person or by a control centre and external commmnication possibilities will be needed besides protection devices.

Pattern recognition has to handle a larger pictorial space with a greater variety of relevant patterns. Hierarchical patternprocessing and decision making may be reasonable.

It has to be taken into account the movement of vehicles relatively to the ground for the mobility. In some cases, a switch from a moving vehicle to ground or vice versa has to be mastered.

Some parts have to be designed as wear out elements with respect to severe conditions of operational life.

Flexibility and adaptability require pattern recognition and the ability to learn even if some conditions may be harmonized by standards.

At present no robot is available, which can be used for such conditions in transportation. Further research and development is needed.

6. Outlook

The market conditions are difficult for a forerunner. The potential users are only forced partly to improve the present situation by robots and the knowledge and skill on such technology is low in this area.

But further development should be in the interest of society, authorities, taxpayers, trade and industry, and customer or passenger.

Moreover the advantage of getting new experience and new solutions can broaden possible application in other areas. A cooperation in research and development for the application of robots in the operational level of transportation is a need:

- for opening the market by achieving world wide standards, by stimulating potential users, and by enforcing the demand for improving the present situation

- for sharing experience of research, users and industry as well as of different fields or regions

- for reducing risk and costs for new developments

- for intensifying political support.

For example, a precompetitional approach would fit to the EURET program (European Research and Development for Transportation) of the DG VII of the Commission of the European Communities. Solutions on the product level could be developed in the scope of EUREKA if only an European level is considered.

But such programs and organizations should not be considered as a permanent set-up.

Working groups with few formalism, but with rigorous efficiency control for participation have been proven very successful especially if they are not only branch oriented rather experts from different fields are embraced. Of course, management and not only administration are needed for achieving progress and for getting problems solution.

7. Bibliography

Genser, R. (1986). Control in transportation systems. IFAC Newsletter, 6, 2-3 Genser, R. (1987). Austrian General Conception of Transportation: A case study. In: Genser, R., et.al. (Ed.). Control in Transportation Systems 1986. Pergamon Press, Oxford, 47-54

Sabine Stifter, J. Lenarčič (eds.)

Advances in Robot Kinematics

With Emphasis on Symbolic Computation

1991. 208 figures. XIV, 484 pages.
Soft cover DM 148,–, öS 1036,–
ISBN 3-211-82302-6

Prices are subject to change without notice

This volume includes a selection of papers presented at the second workshop on Robot Kinematics held in Linz, September 10-12, 1990.

The papers present new results and overviews on various aspects of robot kinematics such as modelling and computation, analysis and design, motion planning and control, inverse kinematics calculations, kinematic redundancy, and parallel mechanisms. Special emphasis was put on the investigation of symbolic computation techniques for problems in robot kinematics.

Springer-Verlag Wien New York

Sachsenplatz 4–6, P.O.Box 89, A-1201 Wien · 175 Fifth Avenue, New York, NY 10010, USA
Heidelberger Platz 3, D-14197 Berlin · 37-3, Hongo 3-chome, Bunkyo-ku, Tokyo 113, Japan

J. Forslin, P. Kopacek (eds.)

Cultural Aspects of Automation

Proceedings of the 1st IFAC Workshop
on Cultural Aspects of Automation,
October 1991, Krems, Austria

1992. 21 figures. VIII, 113 pages.
Soft cover DM 39,–, öS 275,–
ISBN 3-211-82362-X

(Schriftenreihe der Wissenschaftlichen Landesakademie für Niederösterreich)

Prices are subject to change without notice

In October 1991 experts from different research disciplines, like control engineering, systems engineering, sociology, art, philosophy, and politics met in Krems (Austria) to discuss the interplay between recent developments in automation and the culture and social framework, with special emphasis on the approaches in the East and the West. Main topics of these intensive discussions were technology design, automation software and culture, social conditions, education, computer and art, design of man-machine-systems, CIM and culture as well as appropriate methods for interdisciplinary research. A selection of papers presented at this conference can be found in this volume.

Springer-Verlag Wien New York

Sachsenplatz 4–6, P.O.Box 89, A-1201 Wien · 175 Fifth Avenue, New York, NY 10010, USA
Heidelberger Platz 3, D-14197 Berlin · 37-3, Hongo 3-chome, Bunkyo-ku, Tokyo 113, Japan